THE GEOLOGICAL HISTORY OF THE BRITISH ISLES

ARLËNE HUNTER
& GLYNDA EASTERBROOK

The Geological History of the British Isles Course Team

Dave Rothery (*Course Team Chairman*)

Glynda Easterbrook (*Course Manager*)

Arlëne Hunter (*Block Author*)

Other members of the Course Team

Colin Bagshaw (*Reader*)

Gerry Bearman (*Editor*)

Angela Coe (*Reader*)

Sue Dobson (*Graphic Artist*)

Steve Drury (*Reader*)

Sarah Hofton (*Graphic Designer*)

Ray Munns (*Cartographer*)

Val Russell (*Consultant Editor*)

Jane Sheppard (*Graphic Designer*)

Ashea Tambe (*Word Processing*)

The Open University, Walton Hall, Milton Keynes MK7 6AA

First published 2004.

Copyright © 2004 The Open University

All rights reserved. No part of this publication may be reproduced, stored in a retrieval system, transmitted or utilized in any form or by any means, electronic, mechanical, photocopying, recording or otherwise, without either the prior written permission of the publishers or a licence permitting restricted copying issued by the Copyright Licensing Agency Ltd. Details of such licences (for reprographic reproduction) may be obtained from the Copyright Licensing Agency Ltd of 90 Tottenham Court Road, London W1P 0LP.

Edited, designed and typeset by The Open University.

Printed in the United Kingdom by The Alden Group, Oxford.

ISBN 0 7492 0138 X

This text forms part of an Open University course, SXR260 *The Geological History of the British Isles*. Details of this and other Open University courses can be obtained from the Course Reservations and Sales Office, PO Box 724, The Open University, Milton Keynes MK7 6ZS, United Kingdom: tel. (00 44) 1908 653231.

For availability of this or other course components, contact Open University Worldwide Ltd, The Berrill Building, Walton Hall, Milton Keynes MK7 6AA, United Kingdom: tel. (00 44) 1908 858585, fax (00 44) 1908 858787, e-mail ouwenq@open.ac.uk

Alternatively, much useful course information can be obtained from the Open University's website, http://www.open.ac.uk

1.1

Contents

1 **Introduction** — 5

2 **Geological time-scales – a brief review** — 8

3 **A global view of Earth history** — 9

4 **Plate tectonics** — 14
4.1 Introduction — 14
4.2 Revealing past plate tectonic events — 14

5 **The main lithotectonic units of the British Isles** — 23
5.1 Introduction — 23
5.2 Precambrian and Lower Palaeozoic Basement — 23
5.3 Caledonian Orogenic Belt — 26
5.4 Older Cover — 27
5.5 Variscan Orogenic Belt — 27
5.6 Younger Cover — 28
5.7 Summary of Sections 1–5 — 29

6 **The Precambrian and Lower Palaeozoic Basement** — 30
6.1 Introduction — 30
6.2 The Basement — 30
6.3 The northern British Isles — 33
6.4 The southern British Isles — 39
6.5 Summary — 40

7 **The Caledonian Orogenic Belt** — 41
7.1 When did the Caledonian Orogeny occur? — 41
7.2 Understanding the British Caledonides: the northern British Isles — 42
7.3 Understanding the British Caledonides: the southern British Isles — 51
7.4 The site of the Iapetus Suture — 58
7.5 Post-orogenic Caledonian granites — 58
7.6 Summary — 59

8 **The Older Cover** — 61
8.1 Introduction — 61
8.2 Interpreting the Devonian sedimentary environments — 64
8.3 Development of the Devonian highlands — 66
8.4 Igneous activity in the Devonian — 67
8.5 Igneous activity in the Early Carboniferous — 69
8.6 The Carboniferous transgression — 69
8.7 Carboniferous palaeogeography — 76
8.8 Summary — 78

9 **The Variscan Orogenic Belt** — 80
9.1 Introduction — 80
9.2 Recognizing Variscan deformation in the rock record — 81
9.3 Permo-Carboniferous igneous activity in the Variscan Foreland — 91
9.4 Summary — 94

10 **The Younger Cover** — 95
10.1 Introduction – from deserts to glaciers — 95
10.2 Changing sea-levels from the Permian to Cretaceous — 95
10.3 Mesozoic to Tertiary igneous activity — 99
10.4 Tectonic development of the Younger Cover — 101
10.5 Palaeogeography of the Younger Cover — 106
10.6 Summary — 110

11 **The Quaternary Period** — 112
11.1 Introduction — 112
11.2 Quaternary deposits and geological maps — 112
11.3 How can the Quaternary be defined? — 112
11.4 How the Quaternary began — 115
11.5 Glaciation of the British Isles — 117
11.6 Between and beyond the glaciations — 121
11.7 Summary — 123

12 **Objectives for this book** — 124

Answers to Questions — 125

Acknowledgements — 128

Appendix — 129

Glossary — 131

Index — 141

The Course Team wishes to express its thanks to those who contributed to Block 6 of the discontinued *Geology* course S236, which formed the basis for parts of this new text.

We also thank Cynthia Burek for her suggestions on the Quaternary.

This book was originally written to accompany an Open University short course of the same name, SXR260 *The Geological History of the British Isles*. The Science Faculty of the Open University offers a range of interesting courses covering Earth Science topics, including another short course S193 *Fossils and the History of Life* and a longer course S260 *Geology*. All of these courses are available for study either on their own or as part of an Open University degree, and no previous qualifications are required to take them. Further information on these and other Open University courses may be obtained from:

The Open University, Customer Contact Centre, Walton Hall, Milton Keynes MK7 6ZX. Tel: 01908 653231.

Alternatively, go to: http://www.open.ac.uk and click on the 'courses and qualifications' button.

1 Introduction

This book is about the geological history of the British Isles, a remarkable part of the world upon which many of the great events in Earth history have left their mark. In this book we use the term 'British Isles' in its geographical sense, referring to the islands of Great Britain and Ireland and the adjacent lesser isles. The British Isles did not exist as such until comparatively recent times, and the surface environment of the continental crust that now forms this region has undergone dramatic changes during the geological history of the Earth.

Figure 1.1 is an artist's impression of what part of northern England looked like during different periods of the region's geological history. You can see that over a relatively short period of time (~100 million years), the area experienced significant changes in:

- climate (varying from subtropical to arid and semi-arid);
- sea-level (with two **transgressions** onto the land by shallow seas); and
- landmass **relief** (a progressive decrease in the height of the highlands).

In the Lake District, a deep ocean has left a thick sequence of **mudstones**, **shales** and **siltstones** overlying and juxtaposed to a series of **volcanic rocks**. In contrast, the somewhat younger rocks **outcropping** in the Pennines (running down the centre of northern England) were laid down in a warm shallow subtropical sea conducive to the formation of **limestones** containing corals and other marine organisms (Figure 1.1a). At several times across the region, marine conditions were interrupted by sands and silts deposited in river

Figure 1.1 A reconstruction of northern England looking westwards towards the Lake District at different geological times from the Early Carboniferous (foot of page) to the Late Permian (top of page). The progressive northwards drift of the British Isles resulted in rocks from a wide range of environments being formed and later preserved. Much of the Lake District highlands consists of Ordovician volcanic rocks. During the Early Carboniferous (a), the region lay close to the equator, with the low ground covered by a shallow sea in which corals and other marine fauna and flora prospered. As this sea receded, this was replaced by the Late Carboniferous coal swamps (b) and later by the deserts of the Early Permian (c), as the British Isles continued to drift northwards towards the present latitude of the Sahara. By the Late Permian (d), a shallow sea had encroached on this area once again.

deltas, with organic material, which eventually formed **coal**, deposited in coastal swamps (Figure 1.1b). Later still in the Vale of Eden (south of Carlisle), **alluvial fans** and sand dunes indicate a desert environment (Figure 1.1c). This was in turn superseded by a shallow sea, now represented by limestones and **evaporite** deposits (Figure 1.1d), which are found throughout the counties of Cleveland and Durham.

In addition to this sedimentary succession, a number of **igneous extrusive** and **intrusive** episodes have left their mark. Periods of uplift, deformation and **erosion** have also influenced the geological history of northern England and indeed the British Isles as a whole.

Geologists find out about these and similar changes by examining the different types of rock formed throughout the geological record, and comparing them with their modern-day equivalents. A number of climatic, sea-level and topographical changes can be recognized throughout the 3.8 billion years of the geological history of the British Isles. The majority of these relate to the ever-changing position of the Earth's **tectonic plates**.

This book will take you on a geological tour of the British Isles to explore how changes in climate, sea-level and relief can be recognized and understood in the geological record. It will not however, be a simple 'look-see' process. Throughout, you will be asked to use a variety of data to help you interpret how and why different rocks formed, and to use this to identify past environments of formation. Although this book examines the geological history of the British Isles (see Figure 1.2), it is important to remember that many of the events that influenced this geological history are local reflections of much wider global events. Therefore, by unravelling the geological history of the British Isles, an insight can be gained into the geological evolution of the whole Earth.

Throughout this book you will find that geological **Periods** may be prefixed by the terms, 'Early', 'Mid-' and 'Late', or 'Lower', 'Middle' and 'Upper'. The first three are used in the context of time (e.g. a specific age period), whereas the last three refer to successions of rocks. So, for example, Lower Jurassic rocks were formed during the Early Jurassic.

This book falls naturally into two main parts: Sections 1–5 deal with general background information on the processes that have shaped the evolution of the Earth. You should read these Sections carefully, as they will reinforce your understanding of the subject. Sections 6–11 are concerned with the geological history of the British Isles, and not only look at the range of rock types found in each geological Period, but also investigate the environments and conditions under which they formed.

Whilst studying this book, you may come across some unfamiliar terms, so to help with this we have included a short glossary (at the end of the book) of the most important ones. Glossary terms appear in **bold** type on the first occasion they are mentioned.

Throughout, you will also notice grid references (in brackets) after locality names. These have been included so that if you wish to do so you can locate them on the two British Geological Survey Ten Mile Maps (North and South sheets), upon which Figure 1.2 is based. Should you wish to purchase these maps, they can be obtained from the British Geological Survey Sales Desk, British Geological Survey, Keyworth, Nottingham NG12 5GG. Tel: 0115 936 3241 or www.geologyshop.com

Introduction

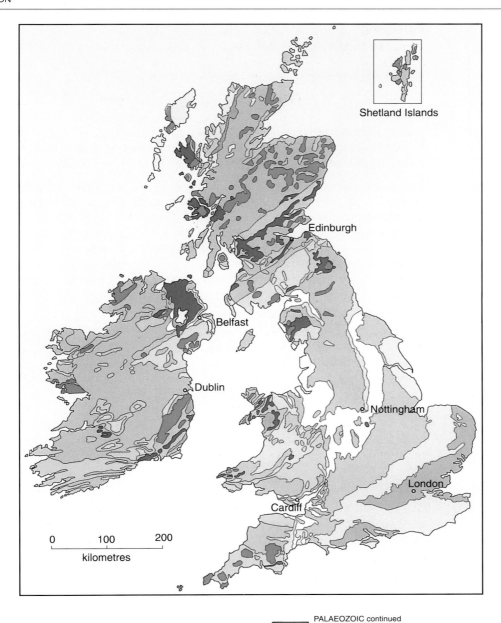

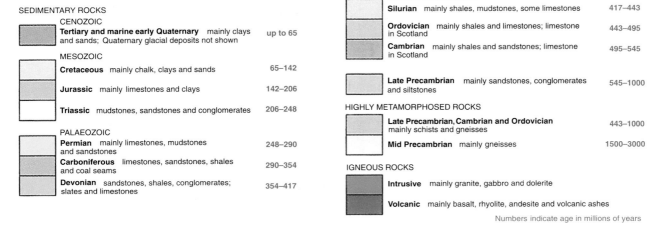

Figure 1.2 Geological map of the British Isles.

2 Geological time-scales – a brief review

Geological time can be divided into a number of **Eons**, **Eras** and Periods, with further subdivisions into sub-Periods or series and epochs. These are arranged chronologically, with the oldest at the bottom, younging upwards to form the stratigraphic column (Figure 2.1).

The **stratigraphic column** can be looked at in two ways. The first deals with the order of rock units. This order has been established using the **Principle of Superposition** and the **Principle of Faunal Succession**, and produces the **lithostratigraphic** ('rock-stratigraphic') **column**, based simply on the *relative ages* for rock successions. The second aspect of the stratigraphic column relates to the geochronological dating of rocks using a variety of radiogenic isotopes. This forms the **chronostratigraphic** ('time-stratigraphic') **column** and allows geologists to apply *absolute ages* to rock successions.

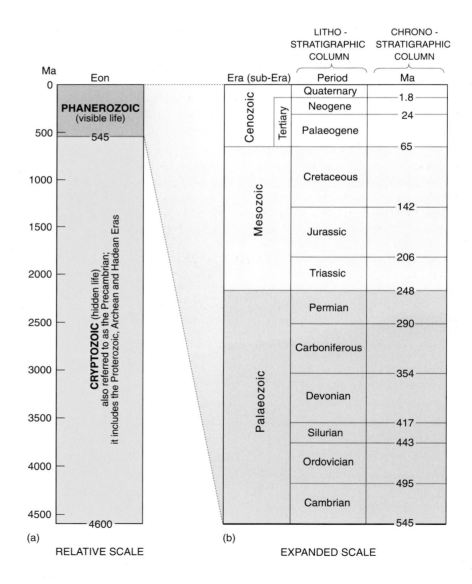

Figure 2.1 The stratigraphic column for the whole of geological time: (a) to true scale and (b) expanded scale for the Phanerozoic. (The Cryptozoic Eon represents ~90% of all geological time.) The scale is divided into Eons, Eras and Periods, which form the lithostratigraphic column. The column of ages of each Period, in millions of years (Ma), forms the chronostratigraphic column. (In 2004, the International Union of Geological Sciences formally named a new Period, to be called the Ediacaran, occupying the final 55 million years of the Cryptozoic Eon.)

3 A GLOBAL VIEW OF EARTH HISTORY

Figure 3.1 shows how the Earth's continents have drifted across the globe over the past 550 million years.

Question 3.1 Describe in one sentence how the geographical position of the British Isles has varied since the Cambrian.

Until the end of the Silurian and beginning of the Devonian, the northern and southern halves of what is now the British Isles were on different continents, separated by an ocean – the Iapetus (Figure 3.1a–c). The existence of this ocean, along with the collision between these continents is recorded by the Caledonian Orogenic Belt, which contains rocks formed within and on the flanks of the now vanished Iapetus (Figure 3.1d). By ~375 Ma (Figure 3.1d–e), this ocean had closed with the resultant continental collision producing a series of major tectonic structures as a result.

At a later date, the Variscan Orogenic Belt (which is found in the southern British Isles) formed as a result of another period of continental collision, when the Rheic Ocean closed between Laurentia and Gondwana (Figure 3.1e). This tectonic activity led to the unification of all the globe's main continental landmasses into one supercontinent called Pangea (Figure 3.1f).

Figure 3.1g and 3.1h shows stages in the break-up of Pangea, which resulted in the formation of new oceans (including the Atlantic), as well as the formation of another extensive **orogenic belt** when Africa collided with Europe to form the Alps, and India collided with Asia to form the Himalayas. These last two examples illustrate that orogenic episodes in one region can occur at the same time as ocean spreading in another region.

In addition to the continental landmasses moving over time, driven by a variety of plate tectonic processes, geologists can also recognize episodic fluctuations in the relative global sea-level throughout the Phanerozoic (Figure 3.2, p.12). Although a detailed study of the causes of relative sea-level change is beyond the scope of this book, it is important to recognize that these processes do occur. If these processes cause the relative sea-level of the whole globe to change, they are referred to as **eustatic** sea-level changes, whereas if they have a more local effect and are due to **isostatic** readjustments (e.g. orogenic movements), they are referred to as **epeirogenic** sea-level changes. In Figure 3.2, the relative change in eustatic sea-level (and mean global temperature) is plotted against time, using the present day sea-level as a baseline.

Figure 3.1 (overleaf)   Reconstruction of continental configurations of the Earth's landmasses during the Phanerozoic Eon. Note how northern and southern parts of the British Isles (red) were dispersed over two continents/tectonic plates until the end of the Devonian (a–d), and that all the landmasses formed one supercontinent during the Permo-Triassic (f).

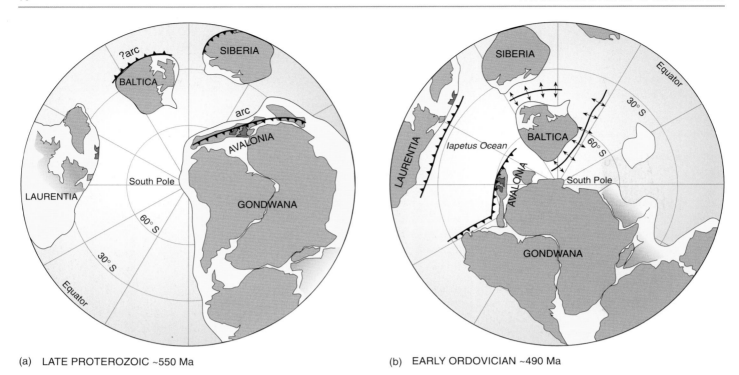

(a) LATE PROTEROZOIC ~550 Ma

(b) EARLY ORDOVICIAN ~490 Ma

(a) The northern British Isles is located at the passive margin of Laurentia, while the southern British Isles is situated behind the subducting margin of Avalonia, a micro-continent on the edge of Gondwana. Both Laurentia and Avalonia are south of ~40° S and are separated from each other by a spreading ocean (which becomes the Iapetus).

(b) The southern British Isles is still located at the margin of Avalonia, which has drifted southwards to ~60° S. In contrast, Laurentia, carrying the northern British Isles, has started to drift northwards, residing at ~20° S, separated from Gondwana by the Iapetus Ocean (which is now beginning to close).

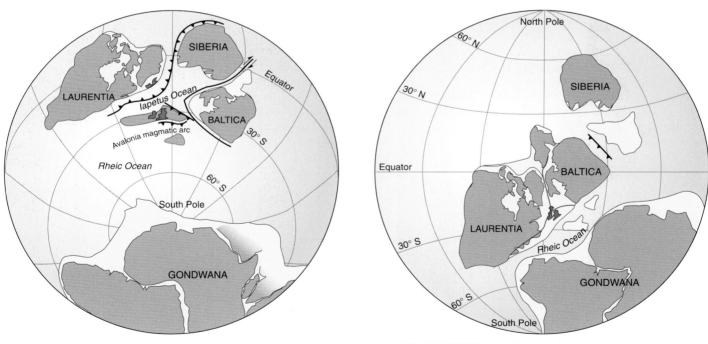

(c) LATE ORDOVICIAN–EARLY SILURIAN ~450–440 Ma

(d) MID–DEVONIAN ~375 Ma

(c) The Iapetus Ocean has been progressively closing, bringing the micro-continent of Avalonia (including the southern British Isles, ~30° S), closer to Laurentia (including the northern British Isles, ~20° S). At the northern margin of the ocean, subduction is occurring below Laurentia, whereas the southern margin with Avalonia is passive. To the south of Avalonia, the Rheic Ocean is actively spreading.

(d) 'Zipper-like' continental collision has been occurring between Laurentia and Avalonia, uniting the British Isles along the Iapetus suture zone (purple line). This collision is known as the Caledonian Orogeny. At this time, the British Isles are at ~20°–25° S, located within the southern desert latitudes.

A global view of Earth history

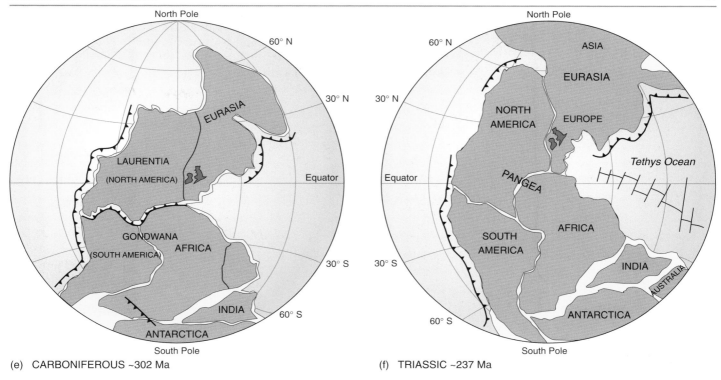

(e) CARBONIFEROUS ~302 Ma

(f) TRIASSIC ~237 Ma

(e) As the Rheic Ocean closes between Laurentia, Eurasia and Gondwana, the Variscan Orogeny starts to affect the southern British Isles. During the Carboniferous, continental drift has carried the British Isles northwards across the equator, into subtropical latitudes.
(f) All of the landmasses have united to form the supercontinent Pangea. To the east, Tethys is actively spreading, while the British Isles continues to drift northwards to 20°–30° N, equivalent to the modern day Sahara latitudes.

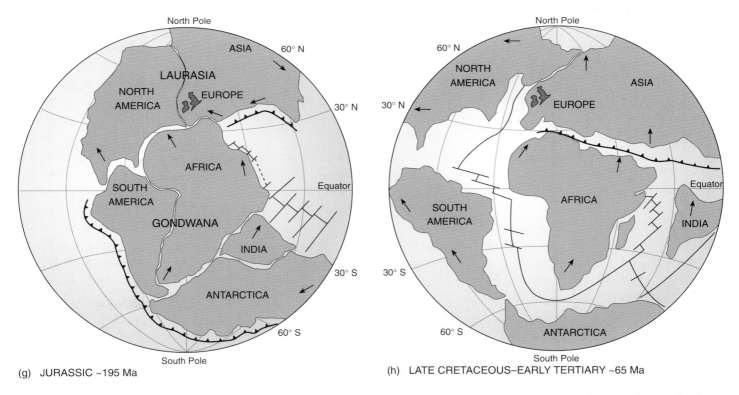

(g) JURASSIC ~195 Ma

(h) LATE CRETACEOUS–EARLY TERTIARY ~65 Ma

(g) Break-up of Pangea results in Gondwana and Laurasia separating, as the southern Atlantic Ocean starts to rift open. The British Isles continues to drift northwards to ~35°–40° N into more temperate conditions, with lithospheric extension and passive rifting occurring to the east (forming the North Sea) and west (where later the North Atlantic will open).
(h) Passive rifting has given way to active rifting to the west of the British Isles, allowing the northern Atlantic Ocean to continue opening in a zipper-like fashion northwards. Active sea-floor spreading is occurring throughout the Atlantic, Indian and Pacific Oceans, whilst Tethys closes, resulting in the eventual collision of Africa, India and Eurasia.

Question 3.2 Compare the sea-level curve in Figure 3.2 with the maps in Figure 3.1 showing the changing assembly of continental masses through time. Briefly describe any correlation between sea-level and the degree of unification of the continental landmasses that you can detect.

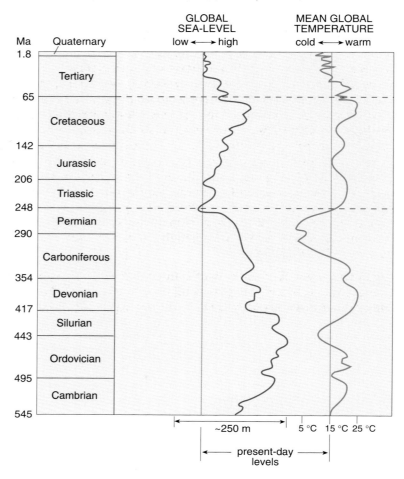

Figure 3.2 Relative global (eustatic) sea-level and mean global temperature changes for the Phanerozoic compared with that of the present day. (Present sea-level and global temperature lines are based on averages for the whole of the Pleistocene, which have been affected by numerous glaciations, lowering the average sea-level and temperature values from today's actual values.)

One cause of global changes in sea-level could be the formation of thick continental ice caps reducing the volume of water in the oceans, resulting in a eustatic fall. When this ice melts, the sea-level must rise. Table 3.1 summarizes the occurrence of major ice ages over the past 2300 million years.

Table 3.1 Occurrence of major ice ages in the geological record and their geographic locations.

Ice age name	Approx. age (Ma)	Geological Period	Geographic areas yielding evidence
Quaternary	0–4	Pliocene–Pleistocene	Many areas including northern Europe
Permo-Carboniferous	270–310	Carboniferous–Permian	Brazil, North Africa
Late Ordovician	450	Ordovician–Silurian	Southern Hemisphere
	600–650		Africa, China, Greenland, Ireland, Scotland, Scandinavia
	750	Precambrian (Proterozoic)	Australia, China, SW Africa
	900		Greenland, Scandinavia, Spitzbergen
	2300	Precambrian (Archean)	Canada, South Africa, USA

Question 3.3   Draw horizontal bars at the appropriate time bands on Figure 3.2 (e.g. 0–4 Ma) to represent the occurrence of ice ages. How well do these data correlate with the global sea-level curve?

What other process can you think of that would affect the sea-level on a global scale?

One clue can be obtained by using Figure 3.1 and looking at what has happened to the continents during periods of eustatic lows. When a supercontinent such as Pangea breaks up (Figure 3.1g–h), a new series of ocean **ridges** will form associated with the formation of new ocean **basins** by **sea-floor spreading**. These ridges can be thought of as submarine mountains that grow in width and length as the spreading process develops. As they grow, the ridges displace water from the ocean basins onto the continental regions. In other words, unlike an ice age, which changes the volume of water *in* the oceans, continental break-up and the formation of new ocean ridges changes the volume *of* the ocean basins, so that seawater is displaced onto the land. Therefore, the sea-level 'highs' of the Early Palaeozoic and Cretaceous may be attributed to an increase in ocean spreading activity. This will be investigated further in Section 10. It has been estimated that the extreme sea-level high during the Cretaceous was ~300 m above present-day sea-level. Not surprisingly, as you will see later, its effects can be seen across the British Isles.

We hope that by now you recognize that the three themes of plate tectonics, climatic changes and sea-level changes are all interlinked. These themes will be revisited throughout this book, as you study the geology of the British Isles.

4 Plate tectonics

4.1 Introduction

The theory of plate tectonics describes three main types of plate boundaries that occur, namely: constructive, destructive and conservative plate boundaries.

- **Constructive plate margins** occur as the result of **lithospheric** extension (Figure 4.1a). As the plates are pulled apart, buoyant mantle wells upwards to prevent a gap from opening. The upwelling mantle undergoes partial melting to form **oceanic crustal** material.
- **Destructive plate margins** form where two plates collide. The more dense plate is subducted below the other, where it is eventually reclaimed by the mantle **asthenosphere** (Figure 4.1b). The location of destructive plate margins is marked by a volcanic arc on the overriding plate, which forms above the point at which partial melting commences. Much of the partial melting is triggered in the mantle wedge by the release of fluids from the subducting slab.
- When two plates slide past each other without creating or destroying crustal material, this is referred to as a **conservative plate margin** (Figure 4.1c). As new crust is not produced at this type of margin, it can be described as being amagmatic. Conservative plate boundaries typically form in oceanic settings and cause constructive plate margins to be offset by tens to hundreds of kilometres, along **strike–slip (transform) faults**.

Different types of igneous processes occur in different tectonic environments, including constructive and destructive plate margins as well as intraplate environments and collision zones. In this Section, you will primarily be looking at how this information can be used to recognize and understand the processes of ocean basin formation and closure in more detail.

4.2 Revealing past plate tectonic events

In Section 3, we referred to the Caledonian and Variscan Orogenic Belts. These are interpreted as representing past destructive plate margins or more strictly speaking, representing the final phases of ocean closure that resulted in continental collision. This Section explores the extent to which it is possible to detect different stages throughout the cycle of ocean basin formation and closure; in other words it examines how geologists can identify:

- the initial rifting and separation of continents;
- the former presence of a wide ocean;
- the closure of an ocean by **subduction**; and
- the collision of two continents once an ocean has closed.

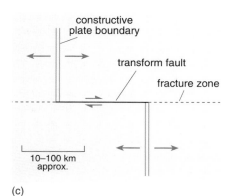

Figure 4.1 Schematic cross-sections through (a) a constructive plate margin and (b) a destructive plate margin (both are drawn to the same scale), along with a sketch map of (c) a conservative plate margin, which in this case, has a transform fault offsetting a constructive plate margin. (Conservative plate margins can occur between any two plates that are sliding past each other without creating or destroying crustal material.)

When doing this, it is important to remember that individual features such as **andesitic lavas**, **granite plutons**, intense folding or high-grade **metamorphism** should *never* by themselves be taken as unequivocal evidence for the occurrence of a particular type of plate margin. It is essential to consider *all* the geological evidence, as well as how all of the evidence fits together on a regional basis. Even if such detective work is successful, it cannot indicate how wide a past ocean was, even though the **suture zone** can often be detected marking where its closure is located. To estimate the width of past oceans, additional information such as **palaeomagnetic** data or **fossil** evidence needs to be used.

4.2.1 Continental extension

Understanding how and why continental plates break apart is extremely important, as this step precedes the formation and development of all new ocean basins. A generalized model for the extension, rifting and separation of continental plates has been developed by examining currently active rifting environments, such as the East African and Red Sea rifts, and comparing these with mature basins, such as the Atlantic Ocean. The East African and Red Sea rifts are regarded as representing continental rifting and the early stages of ocean opening respectively. They can therefore be used as an analogy for how mature oceans, such as the Atlantic or Pacific, first developed. The Red Sea is of particular interest to geologists as a 'living experiment', as the final stages of continental extension and the very beginning of sea-floor spreading can be observed in this one location. At the northern end of the Red Sea, the basin floor consists of thinned continental lithosphere that has been injected by numerous **basaltic dykes**, indicative of the very last stages of continental extension prior to breaking apart. This contrasts with the southern end, where volcanic sites are associated with rifting and the formation of new oceanic lithosphere produced by sea-floor spreading.

Although geologists are primarily interested in the development of new ocean basins, it is important to remember that not all rifts proceed to completion. Many fail after opening just a few kilometres. As you will find out in Sections 6–11, these failed rifts are still extremely important, as they have the potential to be of exceptional economic interest as petroleum reservoirs.

Based on the present-day analogies, a number of stages in the formation, development and eventual closure of ocean basins can be identified (Figure 4.2).

Stage 1: Continental rifting (northern Red Sea stage)

There are two mechanisms for breaking up a continental plate, the simplest of which is to pull it apart under lithospheric extension, forcing the **mantle** to rise up to occupy the 'space' that otherwise would be left by the thinned overlying plate (Figure 4.2a). Continued extension of this already thinned plate will result in it eventually splitting apart. This is often referred to as **passive rifting** (Figure 4.2aii), and is driven by extensional processes. Another way of thinning the continental lithosphere is by stretching it upwards rather than laterally. This can be caused by a hot, buoyant **mantle plume** rising up through the mantle, and forcing the continental lithosphere to dome upwards. This is referred to as **active rifting**, and is driven by mantle processes (Figure 4.2aiii).

With both rifting mechanisms, as the continental lithosphere is stretched, the upper crust undergoes brittle failure forming a series of **horsts** and **grabens**, which progressively separate along normal faults, while the lower crust is subjected to **ductile** stretching and thinning. Although the end result of both passive and active rifting is the same (i.e. the division of a continental plate and the formation of new oceanic crust), the sequences of events leading up to rifting are very different.

Figure 4.2 Schematic cross-section of the sequence of events leading to the formation, development and closure of an ocean basin.
(a) Crustal thinning. At first, the upper parts of the crust extend by developing a series of brittle normal faults. This can be initiated either by passive rifting (aii), or by active rifting (aiii). After the crust extends, it will eventually undergo 'thermal sag', creating a basin in which sedimentary and/or volcanic rocks can accumulate.
(b) Embryonic ocean basin formation. As extension continues, the lower lithosphere will rise and melt.
(c) Active rifting and formation of a passive continental margin.
(d) Onset of subduction.

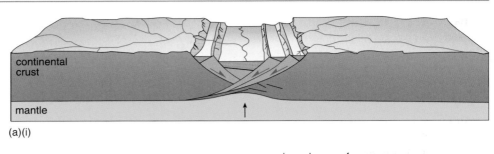
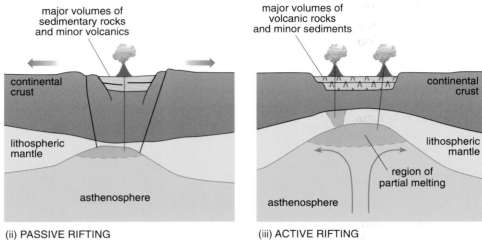
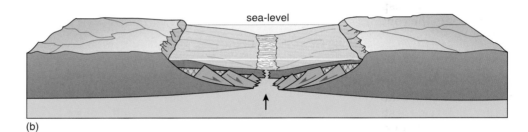
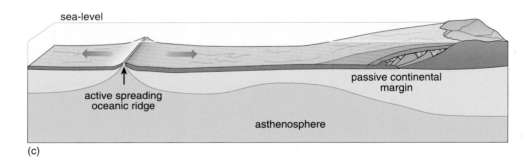
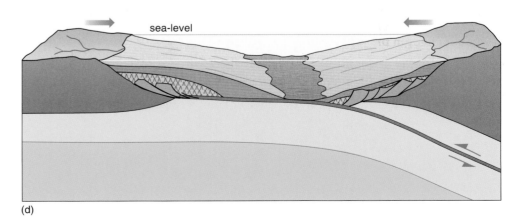

- Based on Figure 4.2a, what do you think will be the first visible effects of passive and active rifting processes on the continental plate?
- With passive rifts, the first visible effect is the formation of a rift zone produced by lithosphere extension. This is followed by regional doming (as hot asthenosphere wells up under the thinned lithosphere), and then by minor volcanism. By contrast in active rifting, regional doming occurs first (as the mantle plume forces the lithosphere upwards), followed by the eruption of huge volumes of extrusive material (e.g. **flood basalts**), ending with the rifting stage.

As you will see in Section 10, the North Atlantic Tertiary Igneous Province, which consists of flood basalts covering areas of Greenland, northern America, Skye and County Antrim, formed because of active rifting that led to the opening of the North Atlantic Ocean. This contrasts with the more minor (but still impressive) Carboniferous volcanic features found throughout the Midland Valley of Scotland (including the Castle Rock and Salisbury Crags in Edinburgh), that formed as a result of passive rifting associated with the Variscan Orogeny (Section 8).

Stage 2: Embryonic ocean basin formation (southern Red Sea stage)

If extension and rifting progress sufficiently, this will lead to the development of an embryonic ocean along the site of the earlier rift zone (Figure 4.2b). Prior to true oceanic lithosphere being produced, basaltic **magma** will be repeatedly intruded into the continental lithosphere along fractures and shear zones. Continued intrusion will eventually lead to the development of a complex of sub-vertical **sheeted dykes**. If these dykes allow magma to be extruded onto the basin floor, **pillow lavas** or small lava flows will result. As the embryonic ocean grows, an igneous sequence typical of sea-floor spreading (i.e. a layered sequence of lavas, sheeted dykes, **gabbro** and **peridotite**) will form. While this embryonic ocean is still narrow, water circulation will be restricted, allowing organic-rich and/or evaporitic sediments to accumulate. Basin subsidence will occur rapidly as the new ocean opens and crustal blocks on either side of the basin are stretched and faulted.

Stage 3: Passive continental margin formation

Eventually, movement along faults initiated during the continental rifting stage ceases, and the entire continental margin starts to subside. Subsidence at this stage occurs because of lithospheric cooling as the distance between the margin and spreading ridge-axis increases, rather than as a result of tectonic movement of the fault blocks (as during the continental rifting stage). By now, all tectonic activity is focused at the new oceanic spreading axis and the continental lithosphere can be referred to as a **passive continental margin** (Figure 4.2c). Throughout this stage, sediments deposited on the passive continental margin **prograde** laterally (advance seawards).

In summary, important indicators of continental rifting and separation include the association of basaltic igneous activity with the development of relatively narrow rift-basins, followed by more widespread regional subsidence. This is not limited exclusively to continental margins, but can also be exhibited by intracontinental rifts that have failed to develop by forming an ocean-spreading centre.

4.2.2 Closure of an ocean

Some of the oldest rocks in the British Isles are over 2500 million years old, whereas elsewhere on the Earth rocks as old as 3800 million years have been found within the continents. These ages contrast with that of the world's oldest *in situ* oceanic crust (excluding **obducted ophiolites**), which is only 200 million years. Considering that basins must have been opening and closing for approximately the same amount of time as continental crust has been forming, this implies that after an ocean basin has grown for a period of ~200 million years, the adjacent passive continental margin becomes a destructive margin and starts subducting the oceanic lithosphere (Figure 4.2d).

❑ Why does this change occur?

■ When the oceanic lithosphere first forms, it is relatively hot and buoyant. As it moves away from the spreading centre and underlying heat source, it cools down and becomes less buoyant.

By the time the oceanic lithosphere is ~200 million years old, it is so cold and dense compared with the adjacent continental lithosphere, that it starts to sink into the underlying asthenosphere under its own weight (helped by the continual push of newly formed oceanic lithosphere). This initiates subduction. If the rates of subduction and sea-floor spreading are equal, this system will remain stable. If however, subduction is occurring at a faster rate than spreading, convergence will occur and may result in collision of the two plates, eventually forming an orogenic belt.

Recent research on how long it takes for a passive margin to switch to an active, destructive margin suggests that this can be as little as ~10 million years. Therefore, a complete cycle of ocean-basin formation, opening and closure should take approximately 400–500 million years (i.e. ~200 million years to open + ~10 million years for the switch from a passive continental to destructive margin + ~200 million years to close).

❑ What evidence do you think geologists can use to identify the presence of past oceans and their margins?

■ They can look for specific igneous, sedimentary and metamorphic assemblages that are characteristic of subduction zones, **continental slopes** and deep ocean basins.

In addition, specific structural features that are characteristic of a subduction zone can be looked for.

Specific examples of evidence for a destructive plate margin are:

Igneous rocks: At subduction zones, **tholeiitic** and **calc-alkaline*** basalts are formed by the release of fluids from the subducting slab into the overlying **mantle wedge**, which undergoes **partial melting** (Figures 4.1b and 4.3a). These basaltic melts rise up into the crust, where they can be temporarily stored in a **magma chamber** and undergo **fractional crystallization** eventually to form andesites and **rhyolites** (extrusive) or their intrusive equivalents, diorites and granites. In general, **island arcs** are dominated by **mafic–intermediate** extrusive rocks, whereas **active continental margins** are dominated by intermediate–

---

*The terms 'tholeiitic' and 'calc-alkaline' (and **'alkaline'** which will be introduced later) can be used to classify different types of igneous rocks according to their geochemistry and/or subtle changes in groundmass mineralogy. Even though these different categories of rocks cannot be distinguished from each other in hand-specimen, once recognized (by detailed petrological and geochemical studies), they can help petrologists to identify the environment of formation. For example, although tholeiitic rocks form in most igneous environments, calc-alkaline rocks are generally restricted to mature arc systems. Alkaline rocks meanwhile represent melts that have formed at deep levels in the upper mantle, and are common at the back of mature island arcs or in continental rifting zones.

felsic plutons. In addition, magmas that are rich in CaO and alkali elements (Na$_2$O, K$_2$O) and have a low FeO/MgO ratio (known as 'calc-alkaline magmas') are indicative of a subduction zone environment, and are typically more abundant at mature (i.e. older, more established) island arcs and active continental margins than immature island arcs.

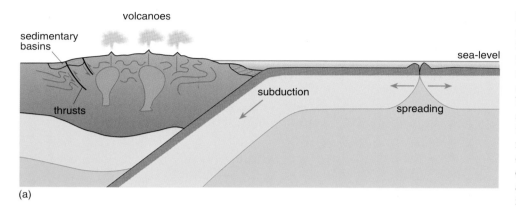
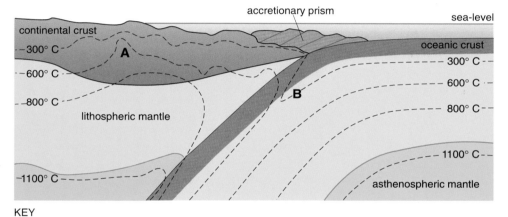
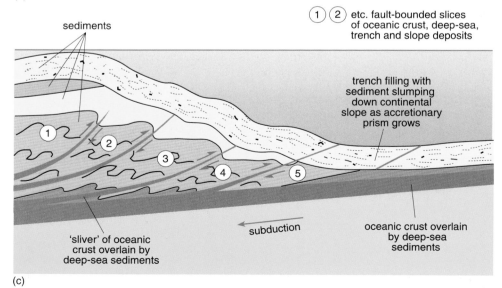

Figure 4.3 (a) Schematic cross-section through a subduction zone involving continental crust, illustrating the key features described in the text. (b) Detailed sketch showing the arrangement of a paired metamorphic belt, which runs parallel to the subducting plate boundary. High-temperature, low-pressure metamorphism (A) occurs in the overlying continental plate, where the elevated geothermal gradients result in a rapid increase in metamorphic facies from greenschist to amphibolite and maybe granulite facies with increasing depth. The low-temperature, high-pressure metamorphism (B) in the subducting oceanic plate is characterized by blueschist facies, indicative of low geothermal gradients.
(c) Detailed sketch of the principal features of an accretionary prism developed in the trench above a subduction zone. A series of thrust-bounded slices of oceanic crust and sediment develops on the margin of the opposing plate, as the downgoing plate carries material beneath the sedimentary-tectonic pile. As the newest slice ⑤ is inserted, it alters the orientation of earlier slices ①–④, so that older fault planes become steeper. Sedimentation is contemporaneous with thrusting, with slumps and turbidity flows triggered by tectonic steepening of the trench slope.

Metamorphic rocks: The key metamorphic indicator of a past subduction zone is a **paired metamorphic belt** (Figure 4.3b). In this belt, the subducting oceanic plate
is subjected to low-temperature, high-pressure (i.e. **blueschist**) metamorphism, while the overriding plate is characterized by high-temperature, low-pressure **greenschist** to **amphibolite** (or even **granulite**) facies metamorphism, associated with the intrusion of magma and thickening of the continental crust.

Sedimentary rocks: The **continental shelves** bordering subduction zones are relatively narrow, with sedimentary material transported rapidly from the source area into the oceanic **trench**, by a series of high-energy density currents called turbidity currents. A **turbidity current** consists of a dense mixture of sediment and water that flows downslope beneath the overlying, less dense, clear water. The end result is a succession of **fining-up** units formed each time the sediment is deposited, initially from the bed load and then out of suspension. Each unit is referred to as a **turbidite**. Turbidites can contain grains from nearby volcanic terrains as well as deep-sea sediments. Turbidite sequences deposited on the downgoing plate are not generally subducted, but accumulate as an **accretionary prism** above the subduction zone (Figure 4.3a–c). (A good turbidite sequence can be seen at Tebay (NY(35)6105) on the edge of the Lake District.)

Structure: Accretionary prisms form in the oceanic trench directly above the shallowest part of the subduction zone (Figure 4.3b–c). In many cases, slices of trench-fill sediments and oceanic crust become detached from the subducting slab and stick to the overriding plate. As subduction continues, these slices of detached sediments and oceanic crust begin to stack up and form a series of **thrusts**, as a result of tectonic accretion. While more slices are tectonically accreted, early-formed thrusts are progressively rotated so that they dip more steeply. With continued sedimentation into the trench, the accretionary prism can become covered by younger turbidite successions, which themselves may be included in later tectonic accretion and thrusting as subduction continues.
The end result of this is that *within* each thrusted slice, the *sedimentary successions get younger upwards and in the direction away from the downgoing plate*, and pass upwards from deep-water sediments (e.g. **chert**, black shales) to trench, trench slope and shelf deposits (Figure 4.3c). However, *the ages of successive thrusted slices decrease towards the trench and downwards in the prism*, with faulting changing from *low angle* near the trench front to *high angle* towards the overriding plate (Figure 4.3c).

You will learn about real examples of these features typical of a past subduction zone and accretionary prism in Section 7, which is concerned with the Lower Palaeozoic rocks of the Southern Uplands, the north of Ireland and northern England.

4.2.3 How wide were the oceans?

Once evidence has been found to prove the existence of an ancient ocean, is it possible to calculate its maximum width? Palaeomagnetic studies can give geologists an idea of the palaeolatitude (N–S) of the ocean but not its palaeolongitude (E–W), so depending on its orientation, an indication of how wide it was may not be possible. However, an approximate indication of how wide the former oceans were can be obtained by examining the fossil **faunal assemblages** that are present (e.g. the range of **species** and type of **biota** present over a specific area). Assemblages on one side of the ocean may differ from those on the other, with the range of species only converging once the ocean has become sufficiently narrow (i.e. closed) for the biota to migrate across the basin. Of course, the opposite of this is that presently separated continental masses that have nearly identical fossil fauna assemblages must once have been united (e.g. southern India, South Africa and Antarctica, which were once united as Gondwana, Figure 3.1).

4.2.4 Continental collision

It should be clear from the above discussions that every ocean basin has a finite lifetime. As the basin closes, former passive and/or destructive plate margins are brought together and eventually collide. This can result in a discrete series of large continental and oceanic crustal fragments being wedged against each other. Collision is often oblique, in which case they are separated by major strike–slip fault systems. Each of these crustal fragments is referred to as an **exotic terrane**, and is recognized by its distinct sedimentary, igneous, metamorphic and structural history compared with that of its eventual neighbours.

Figure 4.4 is a compilation of the range of features that can form when destructive and passive margins collide. The five main features associated with the crustal thickening and shortening produced by collision are:

(a) deformation of pre-existing rock units, producing a series of folds and thrusts inclined towards the suture zone;

(b) formation of regional **nappe** structures, produced by outward gravity-driven sliding related to isostatic uplift of the thickened crust;

(c) high-temperature and high-pressure metamorphism;

(d) crustal melting, which produces collision-related granitic bodies; and

(e) deposition of thick post-collision (post-orogenic) fluvial (e.g. river-deposited) mixed sediments, forming **molasse** successions, which are a poorly sorted mixture of sands and **conglomerates** deposited on the landward side of the suture zone, produced by the rapid erosion of the newly uplifted mountains.

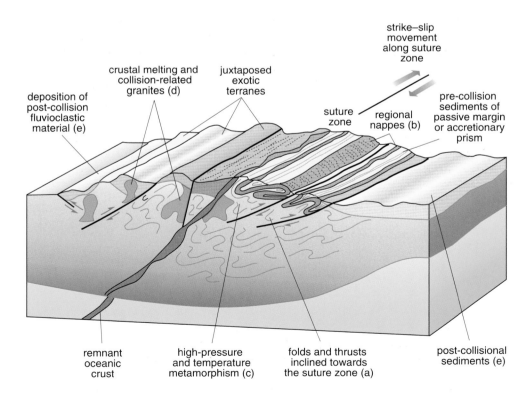

Figure 4.4 Schematic section through a continental collision zone, illustrating the key features discussed in the text.

In addition to the five features listed above, continental collision zones often undergo significant amounts of strike–slip displacement. In some instances, the suture zone is marked by slices of oceanic lithosphere (referred to as an 'ophiolite') that have failed to be subducted, and have instead been forced up on to the other plate (i.e. obducted). Some small and incomplete examples of ophiolites can be found in the British Isles at Ballantrae (NX(25)0882), Fetlar–Shetland (HU(N41)69–HP(N42)61) and on the Lizard peninsula (SW(10)7015).

As you will see when examining the closure of Iapetus (Section 7), the range of rock assemblages present can be used to identify the *types of margin* that formed on either side of a closing ocean, as well as to detect *changes in the tectonic environments* on either side of the ocean, as closure proceeded.

5 The main lithotectonic units of the British Isles

5.1 Introduction

In previous Sections, it was revealed that in the British Isles, the Phanerozoic era was punctuated by two major tectonic periods referred to as the Caledonian and Variscan Orogenies. Both events involved the collision of continental plates, resulting in extensive crustal thickening and isostatic uplift. Over time, the mountain chains formed by these collisions underwent rapid erosion, so that during times of sea-level highs, the sea inundated the land. This allowed new sedimentary units to be deposited, separated from the underlying deformed rocks by a major **unconformity**, representing a significant change in the tectonic and lithological history of the area. Figure 5.1 illustrates the steps involved in the formation of such an unconformity, examples of which can be found marking specific episodes in the geological history of the British Isles.

Using this model, the geological history of the British Isles can be interpreted in terms of a series of distinct orogenic units and their overlying covering units. By doing this, the whole of the British Isles geological history can be simplified into five main units – the Precambrian and Lower Palaeozoic Basement, the Caledonian Orogenic Belt, the Older Cover, the Variscan Orogenic Belt and the Younger Cover (see Appendix at the end of this book). Although these units each have a distinct geological history based on lithology and tectonic structures, they do not correlate with distinct geological periods. Instead, these five units are referred to as **lithotectonic units**.

Note that throughout this Course, **Basement** (with capital 'B') implies the Precambrian and Lower Palaeozoic Basement in the sense used in this Section, whereas **basement** (with lower case 'b') is used to indicate any other rocks underlying a covering lithotectonic unit. For example, the orogenic belt in Figure 5.1 could be referred to as basement, regardless of its age.

5.2 Precambrian and Lower Palaeozoic Basement

The Precambrian and Lower Palaeozoic Basement of the British Isles is a series of nine discrete, exotic terranes whose boundaries are fault systems that have undergone large but usually unknown amounts of lateral and vertical movement over time (Figure 5.2 and Table 5.1). Each **terrane** is a specific geographical area characterized by a distinctive geological history, which was different from that of its current neighbour up until the time that they 'docked' together. Once united, the originally separate terranes then underwent the same geological processes.

Using the rock units present in each area, it is possible to establish the type of environment(s) each terrane originally represented. However, as each terrane is delimited by fault boundaries, it is not always apparent exactly where the terranes originally formed in relation to each other. Despite this we can be sure that the nine terranes that make up the British Isles joined together into their present configuration *between* Proterozoic (Late Precambrian) and Carboniferous (Late Palaeozoic) times.

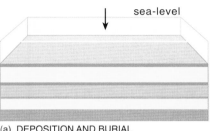
(a) DEPOSITION AND BURIAL

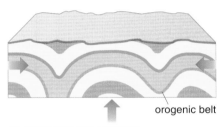
(b) OROGENY AND EROSION

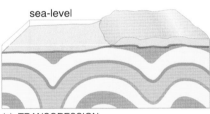
(c) TRANSGRESSION

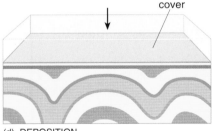
(d) DEPOSITION

Figure 5.1 The development of a major angular unconformity and the relationship between what are termed in the text 'orogeny' and 'cover'. In some instances, the orogenic belt below a cover unit is referred to as 'basement'.

Table 5.1 Summary of the main rock units found in each of the nine discrete, exotic terranes that make up the British Isles.

Terrane	Main rock units
(1) Hebridean Terrane	Archean and Lower Proterozoic gneiss overlain by undeformed sediments from the Upper Proterozoic (terrigenous red sandstones) and Cambrian to Middle Ordovician (shallow-water marine siliciclastics and limestones).
(2) Northern Highlands Terrane	Precambrian to Upper Ordovician and Silurian metamorphosed sediments and mafic to felsic igneous rocks.
(3) Central Highlands (Grampian) Terrane	Upper Proterozoic metasediments overlying gneiss. Based on metamorphic evidence, this terrane is known to have accreted to the previous two by the Early Silurian.
(4) Midland Valley Terrane	A series of small, discrete terranes faulted into one main region, or **superterrane**. The main basement is thought to be an ancient fore-arc region, covered by Upper Palaeozoic volcanics (tholeiitic and calc-alkaline), volcaniclastic sediments, mudstones, sandstones and carbonates.
(5) Southern Uplands Terrane	Lower–Middle Palaeozoic sediments including turbidites, mudstones, pillow lavas, chert and some ophiolite material, indicative of an open, spreading oceanic environment.
(6) Leinster–Lakes Terrane	Lower Palaeozoic marine siliciclastic sediments, mudstones and limestones, along with subduction-related calc-alkaline volcanics.
(7) Monian Terrane	Precambrian calc-alkaline volcanics, blueschists, sediments and **mélange**, representative of a Precambrian to Early Cambrian subduction zone.
(8) Avalon–Midland Platform Terrane	Cambrian–Ordovician sedimentary succession of shallow-water marine siliciclastics and limestones on a Precambrian basement, grading to terrigenous **red beds** and some mafic volcanics.
(9) Lizard Terrane	Middle–Upper Devonian ophiolite complex associated with metamorphosed igneous and sedimentary successions, overlying Ordovician quartzites.

On Figure 5.2, find the thick black line that represents the Moine Thrust Zone. Everything to the west (left) of this line (including the Outer Hebrides and part of the Inner Hebrides) is the Hebridean Terrane.

Much of the Hebridean Terrane is made up from undifferentiated gneiss from the Lewisian complex, which is the oldest rock unit in the British Isles. However, from Rhum (NM(17)3090), up to Enard Bay (NC(29)0513), Torridonian Sandstone is predominant, dating from the Proterozoic (i.e. Late Precambrian).

Also trending across the northern Hebridean Terrane is a series of multiple intermediate to mafic dykes oriented in a north-westerly direction and which also form part of the Lewisian Complex. In this abundance, they can also be referred to as a **dyke swarm**. They do not cut across the Torridonian Sandstone, which indicates that the dykes were intruded *before* the sandstones were deposited.

However, on Skye, there are dykes trending in a NNW–SSE direction, rather than NW–SE. Closer inspection also reveals that they are made up of a different type of rock. In fact, the dykes on Skye are much younger than those to the north, and were intruded as the North Atlantic Ocean began to rift open during the Early Tertiary. These dykes can be followed all the way across southern Scotland into northern England.

The oldest rocks in the southern half of the British Isles consist of Precambrian hornblende **schists** and **gneiss** and mica schists.

Figure 5.2 (opposite) Simplified map of the British Isles summarizing the nine main exotic terranes. Note how the five principal lithotectonic units (see Appendix) are not limited to specific exotic terranes, but occur in several of these across the British Isles. (From Cope, J.C.W., Ingham, J.K. and Rawson, P.F. (eds) (1992) *Atlas of Palaeogeography and Lithofacies*, Geol. Soc. Publishing House.)

The main lithotectonic units of the British Isles

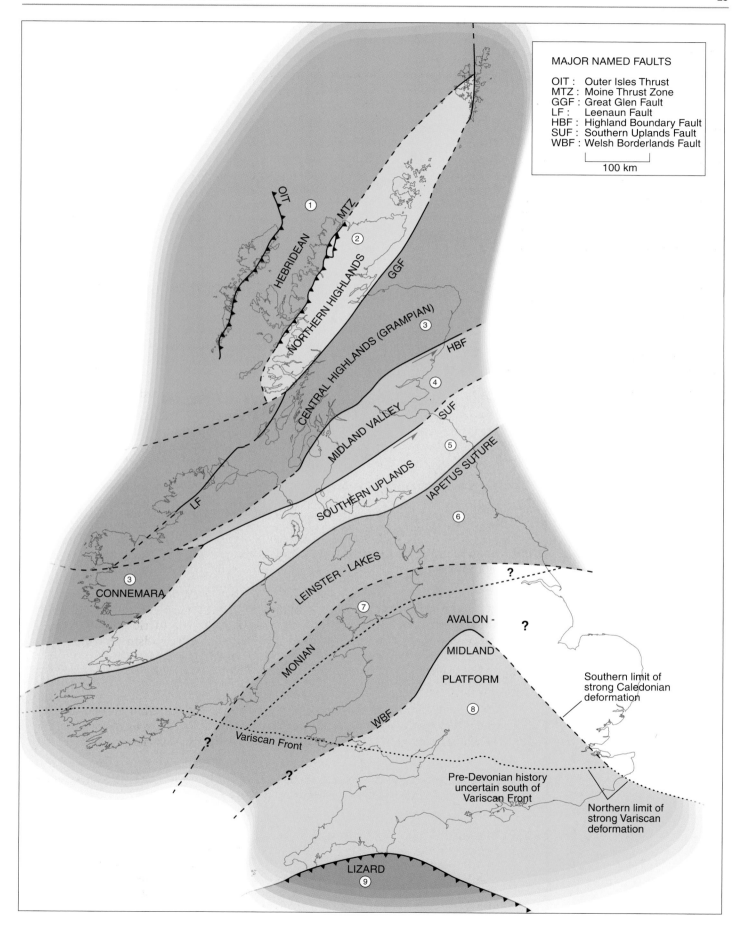

These rocks also occur in Anglesey, north-west Wales (SH(23)48). Similar rocks near the Lizard (SW(10)71) and Start Point (SX(20)83) are now known not to be Precambrian; they are in fact associated with the Variscan Orogeny. There are also intrusive Precambrian rocks in England, the oldest of these occurring near Leicester (SK(43)51). In Ireland, the oldest rocks occur in the south-eastern corner around Rosslare, County Wexford, and are part of the same basement terrane as those in Anglesey (the Monian Terrane, Figure 5.2 and Table 5.1).

The oldest rocks in the southern British Isles are considerably younger than those in the northern British Isles (Figure 1.2). Table 5.1 summarizes the main units found in each of the basement terranes across the British Isles. Although for many purposes the geological history of the British Isles can be described using the simpler five-fold lithotectonic system, it is often helpful to consider these nine discrete exotic terranes, particularly in relation to the structural evolution of the region.

5.3 Caledonian Orogenic Belt

From the Appendix, it can be seen that the Caledonian Orogenic Belt outcrops in several locations across the British Isles including Shetland, northern Scotland (between the Moine Thrust Zone and the Highland Boundary Fault), the north-western and eastern corners of Ireland, the Southern Uplands, the Lake District, the Isle of Man and central and northern Wales.

In the north of Scotland, Shetland and north-west Ireland, the Caledonian Orogenic Belt consists of high-grade metamorphic Moine and Dalradian rocks. This contrasts with the Caledonian Orogenic Belt outcrops from the Southern Uplands southwards, which consist of low-grade or unmetamorphosed successions from the Cambrian to Silurian.

Therefore, the Caledonian Orogenic Belt can be subdivided into the high-grade metamorphic Caledonides (northern Scotland, Shetland and north-west Ireland), and the low-grade to non-metamorphic Caledonides (all locations from the Southern Uplands southwards).

Large granitic intrusions occur in both high-grade metamorphic and low-grade to non-metamorphic parts of the Caledonides in Scotland and northern England, but are absent in Wales (Figure 1.2).

In central and northern Wales, outcrop patterns indicate that major folds have NE–SW trending axes. These are well displayed between Carmarthen (SN(22)4119) and Llandrindod Wells (SO(32)0760), and Montgomery (SO(32)2296) and Llanrwst (SH(23)8061). This fold trend can be followed towards West Dyfed (Pembrokeshire (SN(22)0015)), where it swings round to a more E–W trend.

The same general NE–SW trend can also be seen in the low-grade to non-metamorphic Caledonide terranes of the Southern Uplands and Lake District. This is often referred to as the Caledonian structural trend.

In Anglesey, to the south-west of Bangor and on the Lleyn Peninsula, the Lower Palaeozoic rocks that make up the low-grade to non-metamorphic Caledonides rest unconformably on the Precambrian Basement (some of which is metamorphic).

Throughout the Highlands of Scotland, a general NE–SW Caledonian trend can be recognized, particularly in the Dalradian. Likewise, some major faults with the same trend are also recognized, e.g. the Great Glen Fault, which cuts south-west across the Highlands from Inverness (NH(28)6745) on the Moray Firth, along Loch Ness to Fort William (NN(27)1174) and is picked up again in Northern Ireland at Londonderry ((24)5422) (Figure 5.2). This fault is an important strike–slip fault, the movement of which may have begun during the Caledonian Orogeny, and certainly continued into the Mesozoic.

In Shetland, the trend of the high-grade metamorphic Caledonides swings round to a N–S direction, which can be traced down to the NNE–SSW trend on the Scottish mainland between Findochty (NJ(38)4767) and Portsoy (NJ(38)5766).

The high-grade metamorphic Caledonides and the underlying basement (the Hebridean Terrane) in the north-west of Scotland are separated by several major thrust zones, the main one of which is the Moine Thrust Zone. This dips gently to the south-east. In this area, the Basement consists primarily of the Precambrian metamorphic Lewisian Complex with unmetamorphosed Torridonian sediments, and some Cambrian and Ordovician sediments on top. The Basement has been overthrust from the south-east by the Northern Highlands Terrane, which consists of younger metamorphosed Moine Supergroup sediments. The unmetamorphosed nature of the Torridonian shows that the Northern Highlands Terrane must have been metamorphosed *before* being thrust on top of the Hebridean Terrane.

5.4 OLDER COVER

Moving up succession, the next lithotectonic unit is the Older Cover. Throughout the British Isles, the Older Cover consists of Devonian and Carboniferous strata, with significant outcrops of Carboniferous volcanic rocks occurring in the Midland Valley of Scotland. Minor Carboniferous volcanics also occur in the Derbyshire area around Matlock (SK(43)2560) and Buxton (SK(43)1074).

In many areas, the Older Cover lies unconformably on top of rocks of the Caledonian Orogenic Belt. However, on either side of the Midland Valley, the Older Cover is faulted against older rocks. Between Manchester and Sheffield the Older Cover forms a gentle N–S trending **anticline**. This implies that the major direction of crustal shortening during this folding event was E–W.

5.5 VARISCAN OROGENIC BELT

Unlike the Caledonian Orogenic Belt, outcrops of the Variscan Orogenic Belt are limited to the south-west of England, southern Wales and the south of Ireland (see Appendix and Figure 5.2). However, as you will see in Section 9, this does not mean that the effects of the Variscan Orogeny were limited only to these southern regions.

The Variscan Orogenic Belt, i.e. south of the Variscan Front on Figure 5.2, consists primarily of Devonian and Carboniferous strata, with older metamorphic rocks at Lizard Point and Start Point. In older books and maps, the Lizard Point was incorrectly assigned to the Precambrian. Recent **radiometric dating** suggests that most of the rocks in this terrane were actually formed during the Devonian.

Large outcrops of granite trend from Dartmoor down to the Isles of Scilly (Figure 1.2). Although there are seven discrete granitic outcrops in this area, geophysical surveys have revealed that these are connected at depth, forming one large regional sheet-like plutonic **batholith**. There are no equivalent plutonic bodies in southern Wales or Ireland associated with the Variscan Orogenic Belt.

In south-west England, Devonian strata outcrop as two strips: one along the north coast of Devon, and the other along the south coast of Devon and running into Cornwall. These two strips of Devonian are separated by a belt of Carboniferous strata (Figure 1.2), which forms the axial part of a large synclinal structure. This is actually a complex **syncline** as shown by a variety of smaller-scale folds in the field.

In southern Wales, across the Gower Peninsula, east of Worms Head (SS(21)4087) and in the southernmost part of West Dyfed in Pembrokeshire, north of St. Govan's Head (SR(11)9793), tight folds can be recognized with a WNW–ESE trend.

Although the boundary is shown as a distinct dotted line on Figure 5.2, the tectonic distinction between the Variscan and Older Cover lithotectonic units does not form a sharp well-defined line, but is gradational in nature.

In south-west England, the Variscan is characterized by relatively intense E–W folding, forming a complex syncline. In the Mendips (ST(31)4050 just south of Bristol) and southern Wales, there are a series of **asymmetrical folds**. These intense folds trend E–W in the Mendips, but WNW–ESE in southern Wales. The more open syncline of the south Wales Carboniferous coalfields may be associated with the Variscan folding, but is not included in the orogenic belt on Figure 5.2. To the north of Bristol (ST(31)5070), the fold axes have swung round to a N–S orientation, and are therefore obviously different to the Variscan trend.

It is not possible to detect what lies below the Variscan Orogenic Belt, as the basement is not exposed. A small outcrop of Ordovician rocks near Veryan on the south coast of Cornwall (SW(10)9238) is a small thrust slice and not a stratigraphic **inlier**.

5.6 Younger Cover

The Younger Cover can be found covering a large part of England (Figure 1.2), and to a lesser extent, north-east Ireland, south-west Scotland, Arran, Mull, and the north of Skye. The Younger Cover consists of Permian to Triassic **sandstones**, **breccias**, mudstones and limestones, Jurassic to Cretaceous limestones and carbonate clays and Tertiary to Pleistocene mudstones, sands and clays.

Almost everywhere throughout the British Isles, the base of the Younger Cover (whether Permian or Triassic) lies unconformably over older rocks.

> Question 5.1 Using the contour lines in the Appendix, describe how the thickness of the Younger Cover varies between onshore and offshore areas.

The outcrop of the unconformity below the Younger Cover is highly sinuous. In some places (e.g. near Cheddar (ST(31)4653)) this is because the Older Cover landscape was buried by Younger Cover sediments (in other words this represents a **buried topography**). Elsewhere (e.g. east of Durham (NZ(45)3340)) Younger Cover was deposited over a fairly planar erosion surface and the irregular pattern of the outcrop represents uneven stripping away of the Younger Cover by recent erosion.

Throughout the southern and south-eastern areas of England (through Dorset, Hampshire, Sussex and Essex), there is a series of E–W trending asymmetrical folds. The northern limbs of the anticlines (i.e. the southern limbs of the synclines) are more steeply dipping.

Before leaving this Section, one further lithotectonic unit should be added to the list that is not included in the Appendix. This is the thin cover of Quaternary **drift** that mantles much of the British Isles, deposited between ~2 million and 10 000 years ago. The Quaternary will be examined in detail in Section 11, looking in particular at how it has sculpted and influenced the landscape of the British Isles.

5.7 Summary of Sections 1–5

- A discrete exotic terrane refers to a large crustal fragment that can be recognized by its distinct sedimentary, igneous, metamorphic and structural history compared with that of its eventual neighbours, and has been juxtaposed into position by major strike–slip faults.
- Nine discrete exotic terranes make up the Basement in the British Isles. These consist primarily of Precambrian metamorphosed rocks but also contain some unmetamorphosed sedimentary units from the Lower Palaeozoic (e.g. in north-west Scotland). In general, the oldest Basement rocks in the southern British Isles are considerably younger than those in the northern British Isles. (Sections 5.1–5.2)
- In addition to the nine Basement terranes, the geological history of the British Isles can be interpreted in terms of a series of five distinct orogenic and overlying covering units. These are the Precambrian and Lower Palaeozoic Basement, the Caledonian Orogenic Belt, the Older Cover, the Variscan Orogenic Belt and the Younger Cover. Although each of these units has a distinct geological history, as inferred from its lithology and tectonic structures, they do not correlate with distinct geological periods, and are therefore referred to as lithotectonic units.
- The Caledonian Orogenic Belt consists of the high-grade metamorphic Caledonides to the north of the Highland Boundary Fault, and the low-grade to non-metamorphic Caledonides from the Southern Uplands southwards. In both areas, regional outcrop patterns and fold axes follow the same NE–SW Caledonian structural trend. In the southern British Isles, the Caledonides lie unconformably over the Basement rocks, whereas in north-west Scotland, the Moine Thrust Zone, which is tectonic in origin, separates the Precambrian Basement from the overlying Caledonides. (Section 5.3)
- The Older Cover consists of Devonian and Carboniferous strata that either overlie the underlying Caledonian Orogenic Belt at an angular unconformity, or are faulted up against it. The most significant faults are the NE–SW trending Highland Boundary Fault and the Southern Uplands Fault, which bound the Midland Valley Terrane. (Section 5.4)
- The Variscan Orogenic Belt consists of intensely deformed Upper Palaeozoic rocks and occurs in the far south-west of the southern British Isles. In the south-west of England, the Variscan is intruded by a series of large granitic bodies. (Section 5.5)
- The Younger Cover consists of a succession of weakly folded strata of post-Carboniferous age that rests unconformably on the older lithotectonic units. (Section 5.6)

6 THE PRECAMBRIAN AND LOWER PALAEOZOIC BASEMENT

6.1 INTRODUCTION

We begin by summarizing in Table 6.1 the characteristics and distribution of the main lithotectonic units that make up the British Isles.

Using Table 6.1 as a reference guide, the remainder of this book will focus on the geological evolution of the British Isles. Where appropriate, particular attention will be paid to significant events that have affected the geological history of northern England. Throughout Sections 6-11 you will be referred to Plates 1–17, bound into the centre of the book. These illustrate palaeogeographic reconstructions of the British Isles at various times during their geological history. The Key to these palaeogeographic maps is at the front of the Plates section in the centre of this book.

6.2 THE BASEMENT

- ❏ Before looking in some detail at the Basement of the British Isles, can you recall from Sections 1–5 what the difference between basement with a lower case 'b' and capital 'B' means?
- ■ In general, *Basement* (with a capital 'B') refers to the Precambrian and Lower Palaeozoic Basement, whereas *basement* (with a lower case 'b') is used to indicate any other rocks or rock successions underlying a covering lithotectonic unit. For example, the orogenic belt in Figure 5.1 could be referred to as basement, regardless of its age.

This Section on the Precambrian Basement is quite short, but you should bear in mind that the Cryptozoic Eon covers approximately 88% of the geological history of the Earth (Figure 2.1). Although a lot is known about many of the individual Precambrian outcrops in the British Isles, to understand this Eon fully it is necessary to have a considerable knowledge of what happened elsewhere on the Earth at this time, which is beyond the scope of this book.

In the British Isles, the Basement consists of rocks that are either *older* than those that make up the Caledonian Orogenic Belt (i.e. Precambrian to Early Cambrian in age), or that were left *unscathed* by this event (i.e. unmetamorphosed Lower Palaeozoic sediments; see Appendix and Table 6.1). On this basis, two main geographical Basement groups can be recognized:

- an extensive area in north-west Scotland and Ireland and adjacent offshore regions;
- scattered smaller outcrops in central England, north Wales and south-east Ireland;

As will be seen shortly, this geographical differentiation is paralleled by significant geological differences between these areas.

6.2.1 HOW OLD ARE THE OLDEST ROCKS?

One of the oldest rocks found on the Earth is the Amitsôq Gneiss, a **metasediment** from Greenland dated at ~3800 Ma (Mid Archean, equivalent to the Early Cryptozoic, Figure 2.1).

Table 6.1 Summary of the main characteristics of the five lithotectonic units that make up the British Isles. NBI = northern British Isles; SBI = southern British Isles.

Lithotectonic unit	Geographical location: NBI	Geographical location: SBI	Oldest rock unit and rock type	Boundary with underlying unit	Regional structural trends
Younger Cover	North-west Skye, Mull	Central and south-east England	Post-Carboniferous sediments	Angular discordance	Broadly conformable strata
Variscan Orogenic Belt	N/A but effects of orogeny felt in Midland Valley of Scotland	South-west England and south-west tip of Ireland (effects also felt further north)	Late Palaeozoic highly deformed sediments + granitic plutons	Undetermined	E–W folds, intensity decreases northwards
Older Cover	Orkney, north-east tip of Scotland, Midland Valley, central northern Ireland	North-east England and the Pennines, southern Wales	Devonian–Carboniferous sediments + igneous units	Angular discordance	Major NE–SW faults plus some N–S folds
Caledonian Orogenic Belt	Metamorphic: Shetland, north-central Highlands of Scotland, north-west Ireland	Non-metamorphic: Southern Uplands, Lake District, Isle of Man, east and south-east Ireland, northern Wales	Lower Palaeozoic sediments, metamorphosed north of the Southern Uplands (SU) and unmetamorphosed from the SU southwards	Tectonic (thrust) boundary in NW Scotland; unconformable in southern British Isles	NE–SW regional trend
Precambrian Basement	North-west Scotland, west of MTZ	Sporadic localities in Anglesey, Leicestershire, Rosslare	NBI: Lewisian Complex, undifferentiated gneiss SBI: metamorphic schists and gneiss	N/A	Varied

Question 6.1 What are the oldest rocks in the British Isles and where are they found? (Refer back to Figure 1.2 and Section 5.2 if you can't remember.)

These rocks are considerably younger than those from Greenland and have recently been re-dated using U–Pb **isotopes** to ~2825–2730 Ma (Late Archean, see Figure 2.1). These very ancient rocks are extremely limited in terms of their geographical distribution, and it is not until ~700 Ma (Late Proterozoic, equivalent to the Late Cryptozoic) that sufficient Basement outcrops are found in different parts of the British Isles to allow a systematic Precambrian rock record to be constructed. When discussing the early geological history of the British Isles, it is important to remember that what is actually available for study is a series of individual geographical areas or terranes, each of which is characterized by its own peculiar geological history (see Section 5.2).

Much of what is known about the Earth in the Mid–Late Cryptozoic Eon is based on rocks from the British Isles. To gain an insight into the Earth's geological history prior to ~2800 Ma, **cratonic** (i.e. old and stable) areas of Scandinavia, South Africa, Australia and Canada must be examined. This is especially important for geologists who are involved in the study of how life developed on Earth during its first ~2000 million years.

6.2.2 Development of life on Earth

Primitive life has probably been present on the Earth since ~4000 Ma, initially consisting of very simple, single-celled organisms. Many of these were similar to **cyanobacteria** that lived by photosynthesis and reproduced asexually by single cell division (Figure 6.1). By ~3500 Ma, as well as excreting oxygen into the atmosphere, the descendants of these organisms had started to produce calcium carbonate. This resulted in the formation of **stromatolite** mounds (Figure 6.2a), equivalent examples of which can still be found today in a few sub-tropical locations (Figure 6.2b). The release of oxygen into the atmosphere was a slow process, but it paved the way for more complex oxygen-breathing organisms to evolve. By 2000 Ma, the single-celled organisms began to join up and form cell clusters and chains.

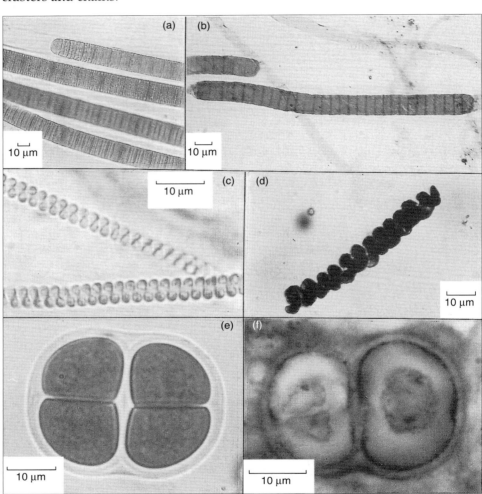

Figure 6.1 Comparison of living cyanobacteria with fossil analogues. (a), (c) and (e) are from stromatolites growing today in Mexico; (b), (d) and (f) are from rocks in the former Soviet Union. (b) is 950 Ma old; (d) is 850 Ma old; and (f) is 1550 Ma old.

As evolution proceeded, cells became increasingly more complex, and by ~700 Ma this had led to the formation of the first **metazoans** (multicellular organisms made up of different organs and tissue types, Figure 6.3). Examples of these early, relatively large, soft-bodied animals were first found in the Ediacara Hills of southern Australia (giving them their name, the 'Ediacara fauna') and in Namibia. Since then, similar examples have been discovered in all of the other Precambrian terranes, including the Charnia Supergroup found in Charnwood Forest near Leicester.

The first skeletal-based organisms appeared in the earliest part of the Cambrian ~545 Ma (Figure 6.4). Their skeletons provided support for soft tissues allowing them to grow even larger and more complex, as well as protecting them from predators.

The Precambrian and Lower Palaeozoic Basement

Figure 6.2 (a) A 3500-Ma-old stromatolite from Western Australia, in cross-section. The specimen is ~20 cm across. (b) Living stromatolites from Shark Bay, Western Australia. Each stromatolite is ~0.5–1.5 m across.

Figure 6.3 A reconstruction of the Ediacara fauna that lived at the end of the Precambrian, just before the Cambrian explosion. The largest organism is ~15 cm long.

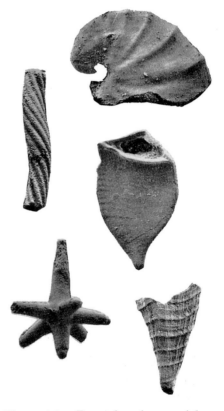

Figure 6.4 Examples of some of the first skeletal-based organisms from the start of the Cambrian explosion. Each of these small shelly fossils is only a few mm long. (From Matthews, S. and Missarzhevsky, V. (1975) *Journal of the Geol. Soc.*, Vol. 131.)

6.3 The northern British Isles

Much of Scotland and western Ireland is underlain by rocks that are either from, or older than the Caledonian Orogeny (see Appendix). This Basement is divided into a series of tectonic blocks, each of which is an exotic terrane.

- ❏ What is the correct definition of a terrane?
- ■ A terrane represents a specific geographical area characterized by a diverse series of rock units, that has undergone depositional, deformation and metamorphic events differing in time or style to those that affected its present neighbours (i.e. it has a distinctive geological history, Section 5.2).

From Table 5.1 and Figure 5.2, it should be apparent that the Basement terranes of Scotland and western Ireland are very different to those in the southern British Isles. However, similarities can be recognized between the Basement of Scotland and western Ireland, and that of northern America and eastern Greenland.

To see how this can be so, look at Figure 3.1a–d. Throughout the Cryptozoic Eon, the northern half of the British Isles was closely associated with northern America and Greenland, forming the ancient continent of Laurentia (Figure 3.1a). At the same time, the southern British Isles was located several thousand kilometres away on the edge of a different sub-continent referred to as Avalonia, which was initially part of the larger continent of Gondwana (Figure 3.1a). By the Mid-Devonian, tectonic drift had resulted in the two halves of the British Isles becoming united, with these lying at the heart of the Caledonian Orogenic Belt, now marked by the Iapetus Suture Zone (Figures 3.1d and 6.5).

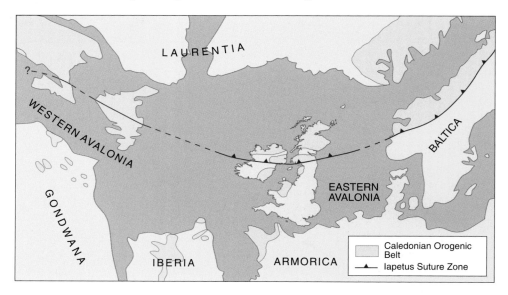

Figure 6.5 Distribution of Caledonian rocks in the present North Atlantic region and the approximate location of the Iapetus Suture Zone. *Note*: Laurentia includes present day Greenland and Labrador. (From Soper, N.J. *et al.* (1992) 'Sinistral transpression and the Silurian closure of Iapetus', *J. Geol. Soc.*, Vol. 147.)

Unravelling the geological history of the Basement of the northern British Isles is not easy, and has led to numerous disputes over the last two centuries. A broad overview of the main events can however be obtained by considering this area as the series of terranes identified in Section 5.2 (Figure 5.2 and Table 5.1). Using this as a model, it can be seen that true Precambrian Basement occurs in the far north-west of Scotland, primarily located west of the Moine Thrust Zone where it is known as the Hebridean Terrane. To the east of the Hebridean Terrane lie the Northern and Central Highlands Terranes (Figure 5.2). These terranes consist of the Moine and Dalradian Supergroups that date from the Caledonian Orogeny, and which are in part underlain by Lewisian Gneiss. These units form the main topic of investigation in Section 7.

6.3.1 The Hebridean Terrane

Question 6.2 Referring back to Section 5.2, what were the main rock units identified in the Hebridean Terrane?

Lewisian Group

The Lewisian Group consists of Archean to Mid-Proterozoic mixed gneiss and mafic–felsic intrusions. The mixed gneiss varies from a coarse to very coarse, pink and white felsic gneiss and is interdigitated with localized zones of intermediate to ultramafic **granodiorites**, amphibolites and **pyroxenites** (Figure 6.6a). Also present are sporadic local exposures of metamorphosed sediments, including **quartzites**, **marbles** and mica or garnet-mica schists.

The Precambrian and Lower Palaeozoic Basement

(a) (b)

Figure 6.6 (a) Typical Lewisian gneiss illustrating the coarse mafic and felsic banding. (Hammer is ~30 cm long for scale.) (b) Cross-cutting Lewisian dykes, north-west Scotland. Near the centre of the photograph, a pale pink felsic dyke can be seen cutting across a slightly darker and narrower dyke. The host rock is dark grey **amphibolite**.

Several large-scale regional and smaller-scale local metamorphic and deformation events have affected the area, so identification of the original Basement rock has not been easy. However, geochemical and isotopic analyses have indicated that it probably consisted of calc-alkaline intermediate igneous rocks.

Question 6.3 What is the most likely setting in which igneous rocks of this nature could have formed?

However, plate tectonics and magma generation may have been rather different in the Archean, so it would be unwise to rely on this evidence. Whatever the setting, isotopic analyses indicate a magmatic age of ≥2900 Ma, with the first metamorphic event occurring some 200 million years later.

Although the Lewisian Group has been affected by numerous metamorphic and deformation events, three major events are recognizable across the whole region. These are the Badcallian, Scourian and Laxfordian events (named after their **type areas** in north-west Scotland). The latter two events were responsible for the dyke swarms identified in Question 6.2 and Figure 6.6b). Figure 6.7 summarizes the sequence of events that combined to form the current day Basement in the northern British Isles.

Torridonian Group

The Torridonian Group is the oldest unmetamorphosed rock unit in the stratigraphic record of the British Isles, and consists of a thick sequence (≤10 km) of reddish-brown shales, **arkoses**, gritstones and conglomerates (Figure 6.8). It was deposited after a considerable time gap of ~700–800 million years (longer than the whole of the Phanerozoic) since the last Lewisian deformation event. During this hiatus (i.e. time gap), the top ~45 km of Basement was eroded away. Geologists know this to be so from petrological studies carried out on Lewisian Gneiss directly below the Torridonian Sandstone, which show that the gneiss must have formed at mid-crustal depths, equivalent to 35–60 km. Therefore, before deposition of the Torridonian sediments, these mid-crustal rocks must have been **exhumed** to form an ancient land surface (Figure 6.7e).

At its base, the Torridonian lies on a very irregular, eroded surface, and buries ancient hills and valleys that had a relief of up to 400 m in places. Considerable thicknesses of sediment continued to be deposited even after all the local basement had been completely buried. The Lewisian Gneiss is only exposed today because some of the overlying Torridonian sediments have been stripped away by later erosion.

Figure 6.7 Cartoon sketch (at a variety of scales) illustrating the progressive formation of the Basement in the northern British Isles from ≥2900 Ma to ~1000 Ma.

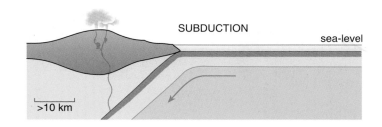

(a) ~2900 Ma. Formation of calc-alkaline crust indicative of a mature island or continental arc.

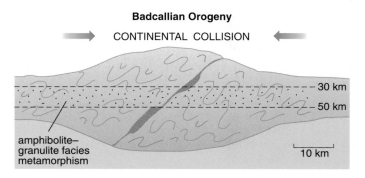

(b) ~2700 Ma. Continental collision buries the calc-alkaline crust to ~50 km depth as part of the Badcallian Orogeny. The associated metamorphism is at amphibolite–granulite grade, occurring at 33–50 km and ~950 °C.

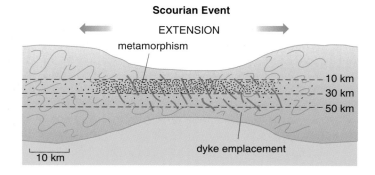

(c) 2400–2200 Ma. As a result of lithospheric extension, the Scourie dyke swarm is intruded at depths of 10–50 km over an extended period of time. Chilled margins along the edges of the basaltic dykes indicate that the Basement must have been at a temperature of <600 °C at the time of intrusion. Metamorphism associated with dyke intrusion occurs at depths of 10–20 km, and is referred to as the Scourian Event.

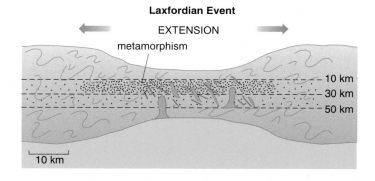

(d) ~1800 Ma. Continued lithospheric extension leads to further intrusion of dykes of intermediate composition, derived directly from the mantle. Associated metamorphism occurs at amphibolite grade (~30 km and ~650 °C), and is referred to as the Laxfordian Event.

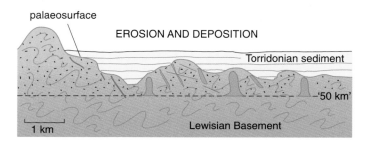

(e) ~1000 Ma. The Basement has been deeply eroded exposing the Scourian and Laxfordian intrusive rocks at the surface. The Torridonian sediments are then deposited on top of an irregular erosion surface, which had a relief of up to 400 m.

Figure 6.8 Changing sedimentary structures found within the Torridonian. In this example from Clare Island, Co. Mayo, the central section consists of a pinching-out segment of cross-stratified medium to coarse sandstone, which is surrounded by a conglomerate that exhibits subtle grain-size variations on a 10–20 cm scale.

Most of the Torridonian consists of planar-bedded units, but examples of poorly-sorted, **graded** and **cross-stratified** successions also occur. The red coloration, sedimentary structures and mixed grain-size of the Torridonian Group indicate that this succession was deposited in a semi-arid fluvial (river) to lake environment. The coarser-grained units were probably deposited by **flash floods**. In addition, the presence of **feldspars** in the Torridonian sediments indicates rapid deposition with very little chemical weathering, to produce a texturally and compositionally immature rock.

The Torridonian is made up of two sedimentary units, dating from 995 ± 17 Ma and 810 Ma, separated by an unconformity. The time gap between the two units is consistent with palaeomagnetic data that reveal a change in palaeolatitude from 15°–33° N for the lower unit to 28°–50° N for the upper unit. These palaeolatitudes are consistent with a semi-arid environment of formation.

Given that the Torridonian buried the entire Basement in the Hebridean Terrane, it is necessary to look further afield for the source of the later Torridonian sediment. Initially, the Torridonian was thought to have been sourced from a more northerly region of the Laurentian landmass currently represented by Greenland (Figures 3.1a and 6.5). However, radiometric studies have shown that the crystallization ages of zircon grains found in the Torridonian sediments are better matched with potential source rocks from Labrador in northern America. It therefore appears that after deposition of the Torridonian Group, the Precambrian Basement of the Hebridean Terrane was displaced from Labrador by **sinistral** strike–slip faulting.

6.3.2 Cambrian–Ordovician sediments

Figure 6.9a illustrates how, after deposition, the Torridonian Group and underlying Lewisian Basement were tilted to the west and subjected to a period of erosion. This formed a planar surface onto which a second group of sediments were deposited during the Cambrian and Ordovician (Figure 6.9b).

Figure 6.9 Schematic sketches (at a variety of scales) illustrating the sequence of tilting and deposition associated with the formation of the Torridonian and Cambro-Ordovician rocks of north-west Scotland.

(a) Sub-aerial deposition of the Torridonian Group on top of the Lewisian Gneiss from 1000–700 Ma. The Lewisian palaeosurface is very irregular and has a relief of up to 400 m in places.

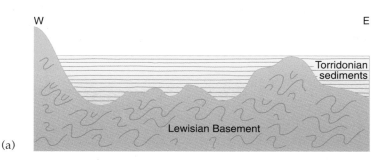

(b) After ~700 Ma, the Lewisian and Torridonian successions were tilted to the west, with erosion forming a new planar surface prior to deposition of the shallow-water marine Cambro-Ordovician siliciclastic and carbonate sediments from ~545 Ma.

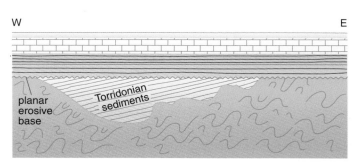

(c) The whole sequence was then tilted to the east, returning the Torridonian sediments to an approximately horizontal position. Some of the overlying Cambro-Ordovician sediments were eroded.

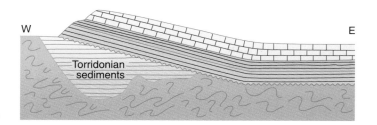

(d) Photograph of Cambrian quartzites (in the foreground) resting unconformably on the Torridonian. The irregular palaeotopography of Lewisian Gneiss is evident in the background of this photograph.

Detailed examination of these Cambro-Ordovician sediments shows that they formed in a marine, shallow-water shelf environment. Over time, the shallow-water marine shelf became starved of **siliciclastic** sediment, so the dominant sediment type changed from sandstones in the Cambrian to carbonate (**calcite** and **dolomite**) muds (Durness Limestone) at the end of the Cambrian and into the Ordovician. The whole Precambrian to Ordovician Basement succession was then tilted back to the east during a second period of tectonism, returning the Torridonian beds to their original horizontal orientation, while tilting the originally flat-lying Cambrian-Ordovician sediments (Figure 6.9c–d).

The Cambrian siliciclastics and Durness Limestone are the youngest lithological units of the Hebridean Terrane. Immediately on top of these unmetamorphosed sediments lies the metamorphosed Moine Supergroup.

The Moine Supergroup was thrust westwards during the Caledonian Orogeny and is separated from the underlying unmetamorphosed sediments by the Moine Thrust Zone (MTZ). Radiometric dates for the Moine Supergroup give an Early Silurian age for metamorphism, implying that a relatively short time (geologically speaking) elapsed between the deposition of the Durness Limestone and overthrusting. Thrusting resulted in the Northern Highlands Terrane docking on top of the Hebridean Terrane about 430 Ma ago. This will be examined in more detail in Section 7.

6.4 The southern British Isles

Several small outcrops of Basement occur in central England, northern Wales and south-east Ireland, consisting of thick sedimentary and volcanic successions along with a few outcrops of gneiss (see Appendix and Figure 5.2). These outcrops are widely scattered and have been subsequently affected by the Caledonian Orogeny, so it is difficult to relate one group to another to build up a palaeogeographic picture of the region. In general terms, the Basement of most of the southern British Isles can be regarded as a superterrane consisting of various terranes that came together at different times (Figure 5.2, Table 5.1). These are:

- the Monian Terrane (not to be confused with the Moinian of Scotland), extending across Anglesey, the western extremities of the Lleyn Peninsula and south-east Ireland, which docked with terranes to the south (shown in the same green colour in Figure 5.2) in the Cambrian;
- the rest of what now became the Avalon–Midland Platform Terrane (the part shown in pink on Figure 5.2), which had become united with the terranes to the north by the Silurian.

Radiogenic ages of between 702 and 549 Ma have been obtained from several volcanic and plutonic rocks from both of these superterranes. This means that they are all Late Proterozoic in age, making the Basement in the southern British Isles much younger than that in the north.

The presence of intermediate to felsic volcanic and intrusive rocks in these terranes points to an active continental margin situated across central England and Wales at this time (Plate 1). Additional geochemical and field evidence for a Late Precambrian–Early Cambrian subduction zone can also be found in Anglesey, which forms part of the Monian Terrane, and which may be associated with a volcanic arc that extended across the adjacent Avalonian Terrane. Not surprisingly, the small amount of available information (limited by restricted outcrop) has led to a number of rival plate-tectonic models.

At present, the current consensus is that from Late Precambrian times, the southern and northern halves of the British Isles were separated by the actively spreading Iapetus Ocean (Figure 3.1a). This continued to open until subduction started along its southern margin (between ~700 Ma and 550 Ma), forming the active continental margin described above and characterized by intermediate to felsic volcanics and intrusives. Remnants of these rocks survive in north Wales and the Lake District. As the rate of subduction exceeded that of ocean spreading, the Iapetus Ocean eventually closed at the end of the Early Palaeozoic, resulting in collision of Avalonia (including the southern British Isles) with Laurentia (including the northern British Isles).

Although the debate about the detailed tectonic interpretation of the Basement rocks of the British Isles continues, there is no doubt that the early geological histories of the northern and southern parts of this region were vastly different. This difference persisted throughout the development of the Caledonian Orogenic Belt, which is the main topic of the next Section.

6.5 Summary

- The Basement in the northern British Isles varies in age from >2900 to ~500 Ma, and consists of a mixture of highly deformed and metamorphosed gneiss, igneous intrusions and sedimentary units, as well as undeformed Lower Palaeozoic sediments.
- After a number of orogenic events ending with the Late Laxfordian Event at 1800 Ma, the Basement remained relatively tectonically stable, until the development of the Moine Thrust Zone at 430 Ma. This event thrust younger metamorphosed successions on top of unmetamorphosed Basement sedimentary units, during the Caledonian Orogeny.
- The Basement in the northern British Isles was relatively unaffected by the Caledonian Orogeny. This implies that rocks from the overthrust, younger Moine Supergroup must have been metamorphosed at some location to the east of the thrust zone *before* being thrust into their current position.
- In the southern British Isles, the Basement is generally much younger than that in the north, ranging from ~700 Ma to 550 Ma.
- In contrast to the north-west of Scotland, which remained a stable cratonic area throughout the Precambrian, the Anglo-Welsh-Irish area contains Late Precambrian andesitic volcanic rocks and granitic intrusions, indicative of a nearby subduction zone. Based on the rock assemblages that make up this succession, it can be stated that the subduction zone was at the edge of an active continental margin.

7 The Caledonian Orogenic Belt

The Caledonian Orogeny is an extremely important part of the geological history of the British Isles, as it was this event that united the majority of the fault-bounded terranes that make up this region. The full extent of the Caledonian Orogeny can be seen on Figure 6.5, with rocks of similar age affected not only across the British Isles, but also along the eastern fringes of North America and the western edges of mainland Europe. Within the British Isles, the Caledonian Orogenic Belt is not limited to a single geographical region but crosses several terranes in Scotland, the north of England, the north of Ireland, and central to northern Wales (see Appendix and Figure 5.2).

> Question 7.1 Using the Appendix as a guide, what tectonic structure forms the north-western border of the Caledonian Orogenic Belt in Scotland?

In contrast to the north-western boundary, the location of the southernmost edge of the Caledonian Orogenic Belt in England and Wales is less certain. This is because the effects of the orogeny decrease progressively southwards across the Leinster–Lakes and Monian Terranes (Figure 5.2), rather than stopping at an abrupt tectonic feature. In Ireland, the southern limit of the Caledonian Orogenic Belt cannot be located at all, as features related to this event have been obscured and overprinted by the later Variscan Orogeny.

7.1 When did the Caledonian Orogeny occur?

The Caledonian Orogenic Belt formed as the result of a number of major global tectonic events that took place during the Early Palaeozoic (Cambrian–Silurian), and extended into the early part of the Late Palaeozoic (Devonian, Figure 3.1b–d). Throughout its development, several periods of crustal extension *as well as* convergence (compression) occurred, affecting both sides of the Iapetus Ocean. Each of these extensional and compressional tectonic events had the potential to influence different terranes by varying amounts of deformation and metamorphism, depending on the terrane's geographical location at that time.

As the terranes on either side of the Caledonian Orogenic Belt were originally separated by the Iapetus Ocean, fossil organisms in the northern and southern sides of the belt are characterized by distinctive and independent evolutionary trends for the whole of the Cambro-Ordovician. It was only in the Late Silurian, when the intervening ocean basin had closed sufficiently, that the fauna on the northern and southern sides of the orogenic belt became mixed. The final period of continental collision, that led to the unification of the British Isles and the eventual end of the Caledonian Orogeny, did not occur until the Late Silurian–Early Devonian.

This complexity of events has led to the term 'Caledonian Orogeny' being restricted to events in the Silurian–Devonian, specifically relating to continental collision. However, all the lithotectonic units involved in the Caledonian Orogeny can be referred to as the British Caledonides, regardless of whether they formed before or during the actual orogenic event.

Before investigating the effects of this continental collision, we ask that you now examine the evolution of the Caledonian Orogenic Belt, looking in particular at how the northern and southern British Isles developed separately throughout the Early Palaeozoic Era.

7.2 Understanding the British Caledonides: the northern British Isles

The Caledonian Orogenic Belt in northern Scotland consists of the Northern and Central Highlands Terranes (located between the Moine Thrust Zone and Highland Boundary Fault). These are made up of Moine and Dalradian metamorphic rock successions (or 'Supergroups') that have been intruded by a series of granites.

A significant amount of fieldwork and associated petrological and geochemical studies have been carried out in this area. This has led to the reassignment of many of the outcrops in the Highlands regions to different stratigraphical units, the results of which are summarized in Figure 7.1a. In simple terms, the Moine Supergroup is believed to be limited to the Northern Highlands Terrane, west of the Great Glen Fault, and is separated from the underlying Lewisian (Precambrian) Basement by a well-defined unconformity. Between the Great Glen Fault and Highland Boundary Fault lies a 'basement group' and the Dalradian Supergroup. This basement group is younger than the Moine Supergroup but older than the Dalradian Supergroup, and may be separated from the Dalradian by an unconformity. The contact between the Moine and the overlying Dalradian Supergroup is even less clear, but has been interpreted as a rift-shoulder unconformity, where the Dalradian strata have been deposited against a rifted margin of Moine rocks (Figure 7.1b).

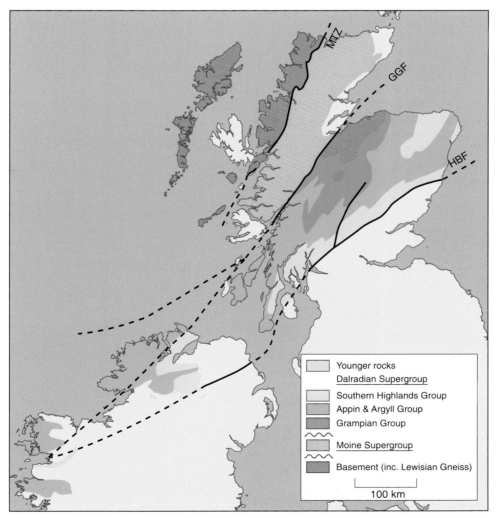

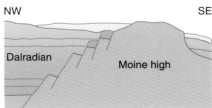

Figure 7.1 (a) Simplified geological map showing the main lithostratigraphic units of the Scottish Caledonides. (b) Reconstruction of the Dalradian Supergroup juxtaposed against the Moine Supergroup at a rift-shoulder margin. (From Tanner, P. W. G. and Bluck, B. J. (1999) 'Current controversies in the Caledonides', *J. Geol. Soc.*, Vol. 156.)

7.2.1 Northern Highlands Terrane – Moine Supergroup

The Moine Supergroup overlies the Lewisian Basement gneiss, and comprises a thick sequence of metamorphosed and intensely deformed Upper Proterozoic sediments. These are conglomerates, sandstones and shales that were originally deposited in a shallow-water marine shelf environment between 1005 Ma and 873 Ma (Figure 7.2a). **Palaeocurrent** directions indicate that sediment transport was towards the north and north-east. After deposition, the Moine Supergroup was intruded by a series of mafic dykes and sheets along with some granitic plutons (Figure 7.2b), before being subjected to two periods of deformation at ~850–800 Ma (Knoydart Orogeny, Late Proterozoic, Figure 7.2c) and ~467–455 Ma (Grampian Orogeny, Ordovician, Figure 7.2d).

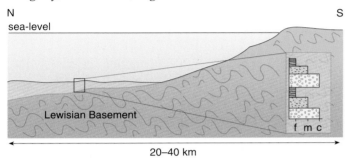

Figure 7.2 Cartoon sequence (at a variety of scales) depicting the formation of the Moine Supergroup.

(a) ~1005–873 Ma. Deposition of conglomerates, sands and muds in a shallow-water marine shelf environment on top of the Lewisian Basement.

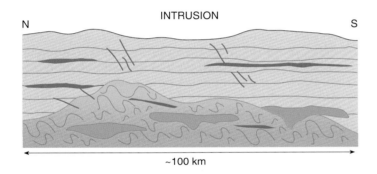

(b) >850 Ma. Period of major igneous intrusion of shallow-level mafic dykes and sheets, and deep-level felsic intrusives into the crust.

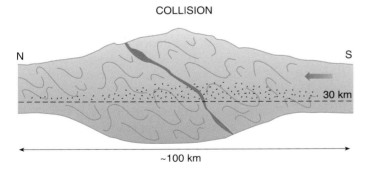

(c) ~850–800 Ma. Continental collision results in lower crustal (~30 km) metamorphism, crustal thickening and large-scale (regional) isoclinal NW–SE trending folding, collectively referred to as the Knoydart Orogeny.

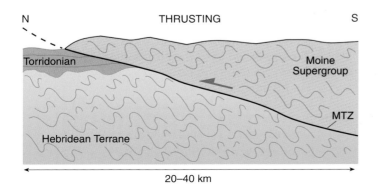

(d) ~467–455 Ma. The onset of the Grampian Orogeny results in the metamorphosed Moine Supergroup being thrust north-westwards over the Hebridean craton, leading to brittle–ductile deformation and the formation of the Moine Thrust Zone.

The Knoydart Orogeny (~850–800 Ma) was a time of regional metamorphism at lower crustal depths of ~30 km and >550 °C, and was associated with a period of major deformation and crustal thickening. This resulted in the formation of large-scale **isoclinal folds** with NW–SE trending axial planes, which frequently contain slices of Lewisian gneiss in their cores. Between the end of this deformation event and the next, the Moine Supergroup did not undergo significant uplift or erosion, but remained at mid-crustal levels.

Between ~467 and 455 Ma, the Grampian Orogeny resulted in thrusting and folding, culminating in the formation of the Moine Thrust Zone (as well as numerous other smaller thrusts throughout the northern Highlands), and the docking of the Northern Highlands Terrane with the Hebridean Terrane. The tectonic history of the Moine can therefore be regarded as one of regional deformation and metamorphism, associated with periods of intensive crustal shortening and thickening.

7.2.2 Central Highlands (Grampian) Terrane – Dalradian Supergroup

The Dalradian Supergroup occurs predominantly in the Central Highlands Terrane (Figures 5.2 and 7.1a), and consists of Late Proterozoic (~850 Ma) to Cambrian (517–509 Ma) sedimentary and igneous strata. The whole sequence is extremely thick (≤ 4 km), comprising sandstones, shales and limestones that have been metamorphosed to schists, slates and marbles, and tholeiitic basaltic volcanics indicative of a rifting environment (Figure 7.3). The Dalradian Supergroup has been subdivided into four lithostratigraphic groups – the Grampian, Appin, Argyll and Southern Highlands Groups (Figures 7.1a and 7.3), with a continuous lithological, structural and metamorphic transition evident throughout all four. Within the Dalradian Supergroup, the grade of metamorphism decreases spatially from the Great Glen Fault to the Highland Boundary Fault, as well as temporally, decreasing from the base to the top of the sequence (Figure 7.1a).

Figure 7.3 summarizes the main stratigraphic and tectonic events involved in the evolution of the Dalradian Supergroup. In simple terms, the lithotectonic evolution of this group is directly comparable with Figure 4.2, and the opening and subsequent start of closure of an ocean basin. The main stages of development are:

(1) *Crustal extension and thinning* (Grampian and Appin Groups, ≤ 850 Ma): formation of a large, subsiding sedimentary basin and deposition of limestones and algal colonies (e.g. stromatolites, Figure 6.2), shales and cross-stratified sandstones resting unconformably on Precambrian Basement, indicative of a shallowing shelf environment (Figures 4.2aii and 7.3a–b).

(2) *Initiation of rifting with continued extension* (Argyll and Southern Highlands Groups, 670–595 Ma): continued lithospheric extension resulted in the sedimentary basins breaking into a series of subsiding fault-bounded blocks, during which time the high rate of extension allowed the intrusion of tholeiitic basaltic dykes and sills (Figure 7.3c–d). This is equivalent to the rifting stage in Section 4.2.1 (Figure 4.2b).

(3) *Completion of rifting, initiation of sea-floor spreading and the formation of a passive continental margin* (Tayvallich Volcanics, Southern Highlands Group, <595 Ma): rifting in the Central Highlands Terrane failed after the eruption of the Tayvallich Volcanics. However, proper sea-floor spreading did occur to the south-east, to eventually form the Iapetus Ocean. Meanwhile, the Central Highlands Terrane became a passive continental margin and the Tayvallich Volcanics became covered by coarse sandstones, resulting in sediment loading and shelf subsidence, followed by cyclical deposition and subsidence of shallow-water marine limestones and shales, grading up into sandstones (Figures 4.2c and 7.3d–e).

Figure 7.3 Cartoon sequence (at a variety of scales), depicting the formation of the Dalradian Supergroup.

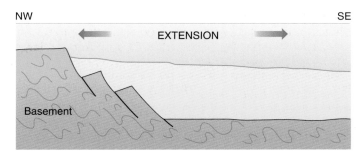

(a) Grampian Group (>850 Ma). Deposition of turbiditic muds and silts into an extending rift-bounded intracontinental basin. The sedimentary succession is indicative of deep-water conditions.

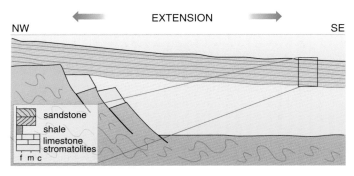

(b) Appin Group (≤850 Ma). Deposition of limestones and stromatolites, shales and cross-stratified sandstones, indicative of a shallowing marine shelf environment. This shallow-water marine shelf built out slowly in a south-eastward direction, away from the continental landmass.

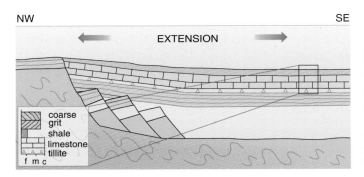

(c) Argyll Group (~670–600 Ma). The base of this group is marked by the deposition of a **tillite** sequence (~668 ± 7 Ma, 1–750 m thick), indicative of a glaciation period at the end of the Early Dalradian. This is directly overlain by shallow warm-water carbonates, followed by coarse, cross-stratified grits, indicative of shallow tidal waters. Due to sediment loading and continued crustal extension, the shallow-water shelf broke into a series of fault-bounded blocks and basins. Sedimentation continued in the form of a series of shallowing-up cycles of limestone, shales and coarse sands and grits.

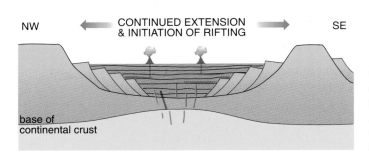

(d) Southern Highlands Group (<595 Ma). Continued crustal extension and basin subsidence eventually resulted in the intrusion of basaltic dykes and sills, indicative of a rifting stage. As extension continued, the magmatic activity climaxed with the eruption of the Tayvallich Volcanics (595 ± 4 Ma), into a series of small rift-basins.

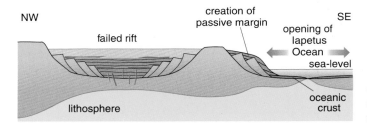

(e) The rift associated with the Tayvallich Volcanics failed to open. However, sea-floor spreading occurred further to the south-east, resulting in the opening of the Iapetus Ocean.

7.2.3 Caledonian deformation of the Dalradian Supergroup

As with the whole of the Moine Supergroup, the oldest Dalradian units (which are adjacent to the Great Glen Fault), have been affected by the Knoydart (~850–800 Ma) and Grampian (467–455 Ma) Orogenies. In contrast, radiometric dates of 630–590 Ma for a range of Upper Dalradian sediments, basaltic lavas and granites (e.g. Tayvallich Volcanics, 595 Ma and Ben Vuirich Granite, 590 Ma) situated next to the Highland Boundary Fault indicate that they are too young to have been affected by the *earlier* of the two orogenies. The Upper Dalradian was, however, deformed and metamorphosed during the Grampian Orogeny (475–463 Ma) and intruded by mafic–felsic igneous complexes, dated at 470–460 Ma.

Within the Central Highlands Terrane, the Grampian Orogeny is marked by major NW–SE compression. This was associated with the collision of a small volcanic arc (possibly part of the Midland Valley Terrane) with the southern margin of Laurentia and the start of closure of the Iapetus Ocean.

Although no evidence exists to prove whether or not the Northern and Central Highlands Terranes were already locked together along the Great Glen Fault by ~850–800 Ma (allowing the Knoydart Orogeny to affect both terranes), later metamorphic evidence proves that by the Early Silurian, the Hebridean, Northern Highlands and Central Highlands Terranes had 'docked' to form one superterrane. Therefore between ~595 Ma and 463 Ma, the Central Highlands Terrane changed from being a passive continental margin (Section 7.2.2) to an orogenic belt, associated with the start of closure of the Iapetus Ocean.

Directly south of the Central Highlands Terrane is the Midland Valley Terrane (Figure 5.2). Although some small outcrops of Lower Palaeozoic rocks do occur there, they do not match up with the Lower Palaeozoic outcrops in terranes further north or south. This is one of the most exotic but poorly exposed terranes of the British Isles. We will therefore move straight on to the more extensive Lower Palaeozoic outcrops in the Southern Uplands Terrane that can be correlated with other strata of similar age.

7.2.4 Early Palaeozoic sedimentation in the Southern Uplands Terrane

The Southern Uplands Terrane can be divided into three broad bands: Ashgill and Caradoc (Upper Ordovician), Llandovery (Lower Silurian) and Wenlock (Middle Silurian). From this it can be seen that the rocks generally become younger south-eastwards across the Southern Uplands. These sedimentary units have been intruded by large granitic bodies and intermediate–felsic dykes with a predominantly NE–SW trend, the significance of which will be examined later.

The boundary between Ashgill and Caradoc and Llandovery is actually quite complex, with numerous sharp kinks. In addition, there are a number of complex small-scale areas with diamond-shaped inliers and **outliers**. These features, along with the sharp kinks between Ashgill and Caradoc and Llandovery, are the result of isoclinal folding. The lithological units in the Southern Uplands get younger in a south-east direction (just as they do in the various Highlands Terranes, Figure 7.1a).

THE CALEDONIAN OROGENIC BELT

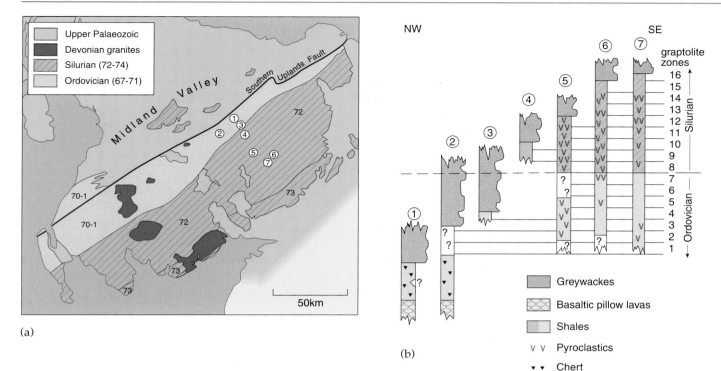

Figure 7.4 (a) Simplified geological map of the Southern Uplands Terrane. Circled numbers ①–⑦ represent the location of the boreholes in Figure 7.4b. The relative positions on the continental slope of boreholes ①,③,⑤ and ⑦ are shown in Figure 7.6. (b) Schematic representation of the successions found in the series of boreholes across the Southern Uplands. Note that the vertical scale represents time and not unit thickness.

The borehole logs in Figure 7.4b are typical of the lithological successions found across the Southern Uplands, with basaltic pillow lavas and **pyroclastics** overlain by relatively thin cherts (fine-grained beds of amorphous silica) and/or black shales rich in **graptolites** (Figure 7.5a). These are followed by extremely thick successions of turbidite deposits consisting predominantly of **greywackes**. Greywacke is a term used to describe a texturally and compositionally immature rock that consists of sand to gravel-sized quartz and feldspar grains as well as rock fragments. These are held together in a muddy matrix that represents at least 15% of the total rock volume. Close examination of the turbidites reveals that they consist of repeated fining-up successions, ranging from sandstones to laminated mudstones (Figure 7.5b). A section through similar turbidite deposits occurs at Tebay (NY(35)6105), which is situated on the edge of the Lake District.

Question 7.2 What environment(s) of formation does this lithological sequence (basalt, chert, shale and turbidites) suggest?

Detailed mapping of rock exposures across the Southern Uplands has shown that as well as being tightly folded, the region is also cut by a number of large faults. These have formed a series of thrust slices in which the sedimentary sequence in each block is significantly different in either lithology and/or age from that of its neighbours (e.g. compare boreholes ①–⑦, Figure 7.4b). You should note that the vertical scale on Figure 7.4b represents time and *not* unit thickness. Each graptolite zone is estimated to span ~3 million years based on the fact that the Late Ordovician and Early Silurian are equivalent to a total time span of ~50 million years. As far as the thickness of each rock type is concerned, the greywacke can be up to several kilometres thick, whereas the cherts, shales and pyroclastic volcanic rocks are at the most 80 m thick.

In general, *within* each thrust slice the strata dip steeply to the north-west and become younger in the same direction (some isoclinal folding may also be present). This means that the direction in which the strata in each block get younger is the *opposite* of that observed across the Southern Uplands Terrane as a whole. This difference in younging direction is an important clue to deciphering how the succession was formed.

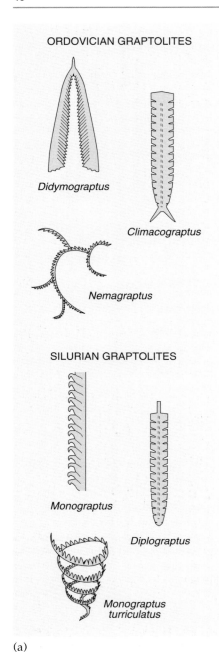

Figure 7.5 (a) Examples of some of the different types of Ordovician and Silurian graptolites found across the Southern Uplands. Graptolites were colonial animals that were probably free-floating suspension feeders, related to Chordata. Each graptolite consists of one or more branches (stipes), on which individual animals (zooids) lived in small cups arranged in one or two rows. The morphology of the graptolites evolved quickly, and they were widespread, so they are ideal **zone fossils**, limited to specific, short periods of geological time.
(b) Part of a turbidite succession similar to that seen at Tebay, with each depositional unit exhibiting a clear erosive base and fining up from light grey fine sandstone to mid–dark grey laminated siltstone and mudstone. To the left of the pencil, some **load casts** can be seen resulting in the light grey sandstone pressing down into the finer-grained top of the previous depositional unit.
(c) Artist's representation of a turbiditic flow progressing down a continental slope.

Question 7.3 How does the age of the greywacke sequence differ *between* thrust slices ①–⑦ on Figure 7.4b? Is there any systematic change in age from the north-west to south-east?

Based on the type of relationship formed, it seems logical to conclude that the greywacke sequence in each thrust slice was originally deposited some distance from the others. Each thrust slice therefore represents an area that is progressively more distant from the continental landmass (and source of sediment) than its north-western neighbour (Figure 7.6).

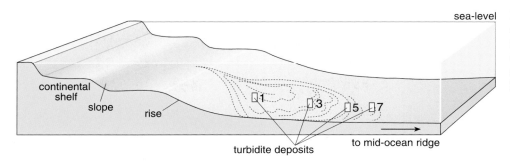

Figure 7.6 Reconstruction of the continental slope and ocean basin, illustrating the original location of the thrust slice sedimentary successions identified from the boreholes in Figure 7.4b.

Question 7.4 To help you visualize this sequence more clearly, complete Figure 7.7 by marking on the regional and thrust slice younging directions (by means of arrowheads), as well as completing the strata ornamentation as shown in the key. (Blocks 1, 5 and 6 have already been completed for you as a guide.)

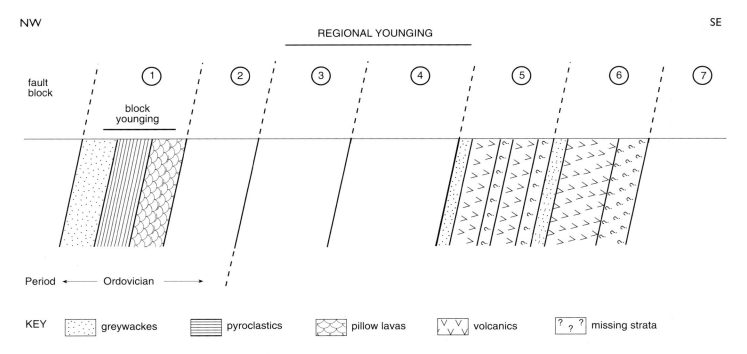

Figure 7.7 Simplified cross-section of the Southern Uplands, across the thrust slices ①–⑦ shown in Figure 7.4. (For use with Question 7.4.)

Question 7.5 What does this assemblage of sedimentary and tectonic features represent? (Hint: look back at Section 4.2.2 and Figure 4.3c.)

This lithotectonic evidence and the model from Section 4.2.2 can be used to put the Ordovician and Silurian sediments of the Southern Uplands into their palaeogeographic context.

During the Grampian Orogeny in the Mid-Ordovician (467–455 Ma), a small volcanic arc collided with the Laurentian landmass resulting in major NW–SE convergence across the Northern and Central Highlands Terranes (see Section 7.2.3). To the south of this arc across the present day Southern Uplands, the Ordovician to Silurian black graptolite shales, cherts and greywackes were progressively thrust into an accretionary prism as the Iapetus Ocean closed (Figure 7.8). The northern margin of this prism is represented by the Southern Uplands Fault. As thrusting continued, earlier thrust blocks in the prism were rotated, resulting in a progressive increase in thrust steepness to the north (Figure 7.8b). Exposures of Silurian rocks in the Hagshaw Hills, ~30 km south of Glasgow (NS(26)6535), reveal that as the accretionary prism developed, parts of it rose above sea-level (Figure 7.8b), like modern-day Barbados, shedding sediments to the north and south.

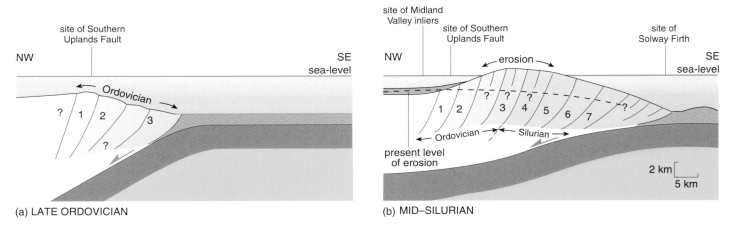

Figure 7.8 (a) Accretionary prism model for the Ordovician–Silurian in the Southern Uplands.
(b) Sub-aerial erosion of the accretionary prism, as the thrust angle increased due to continued collision.

Around Ballantrae (NX(25)0982), the Ordovician strata consist of greywacke successions similar to those in the Southern Uplands, and are underlain by mid-oceanic ridge (**MORB**)-like basalts and pyroclastics (including pillow lavas, Figure 7.9), gabbros and peridotite.

Figure 7.9 Example of some inverted pillow lavas at Ballantrae. The pillows have been inverted as a result of the isoclinal folding and thrusting that has affected this area.

Question 7.6 What does this association of sedimentary and igneous rocks represent?

Before examining what happened to the British Isles when Avalonia finally collided with Laurentia during the Late Silurian to Early Devonian, we will look at the Early Palaeozoic lithotectonic evolution of the southern British Isles. Particular attention will be paid to how and why events here differed from those in the northern British Isles.

7.3 Understanding the British Caledonides: the southern British Isles

7.3.1 Review of the Basement palaeogeography of the southern British Isles

During the Precambrian, the southern British Isles was located at the northern edge of an active continental margin on the micro-continent of Avalonia, ~60° S (Figure 3.1a), and was characterized by sporadic and discontinuous igneous and tectonic activity. Although Avalonia was initially attached to the continent of Gondwana, by the Early–Mid-Cambrian (545–518 Ma), it broke away and started to drift northwards (Figure 3.1b). This northwards drift resulted in the Iapetus Ocean between Laurentia and Avalonia starting to close. Meanwhile to the south, the Rheic Ocean began to open between Avalonia and Gondwana. By the Late Ordovician, subduction was no longer occurring below Avalonia, changing the northern edge of this micro-continent into a passive continental margin (Figure 3.1c). However, subduction was still occurring under the southern margin of Laurentia, so the Iapetus continued to close.

7.3.2 Early Palaeozoic sedimentation in the southern British Isles

Early Palaeozoic exposures are scarce in the southern British Isles, but sediments deposited at this time in England and southern Ireland have been found resting unconformably on the Precambrian Basement. This contrasts with northern Wales, where Cambrian sediments have a conformable relationship with the Basement.

In general terms, variations in Cambrian sedimentation rates and lithologies were predominantly controlled by global changes in sea-level (Figure 3.2). From the Late Precambrian, the global sea-level rose in a series of pulses reaching its highest level during the Late Cambrian. At the start of the Ordovician, global conditions changed, with a dramatic drop in global sea-level, followed once more by a progressive rise. Until recently, the general rise in global sea-level that began in the Late Precambrian was attributed to the fragmentation of the Precambrian landmasses and formation of young, hot oceanic ridge systems (Figure 3.1a–b), which displaced the oceans onto the land. It is now believed, however, that changes in the mean global temperature also played a significant role in controlling the global sea-level (Figure 3.2 and Table 3.1). At the end of the Precambrian, the Earth switched from being a global **icehouse*** (the climax of which formed the **tillite** sequence dated at ~670 Ma and found in the Dalradian Supergroup, Figure 7.3c), to a global **greenhouse*** in the Early Palaeozoic. The pulsed sea-level rises that occurred throughout the Cambrian to Early Ordovician can therefore be at least partially attributed to a fluctuating ice cap over the southern pole, with each rise corresponding to a warming stage.

Associated with these increases in sea-levels and periods of global warming, was a change in the type of sediment being deposited. Great thicknesses of evaporites and shallow-water carbonaceous muds formed as the oceans flooded across the shallow continental areas. The new shallow-water environments suitable for colonization led to a massive diversification in marine fauna.

The remainder of this Section in association with Figure 7.10 and Plate 2 summarizes the main palaeogeographic features and lithologic units formed across the southern British Isles during the Early Palaeozoic.

---

* 'Icehouse' and 'greenhouse' can be applied to periods in the Earth's geological history when mean global temperature was respectively lower or higher than mean temperature today. Associated with these changes in global temperature is the spread or retreat of major glacial ice caps.

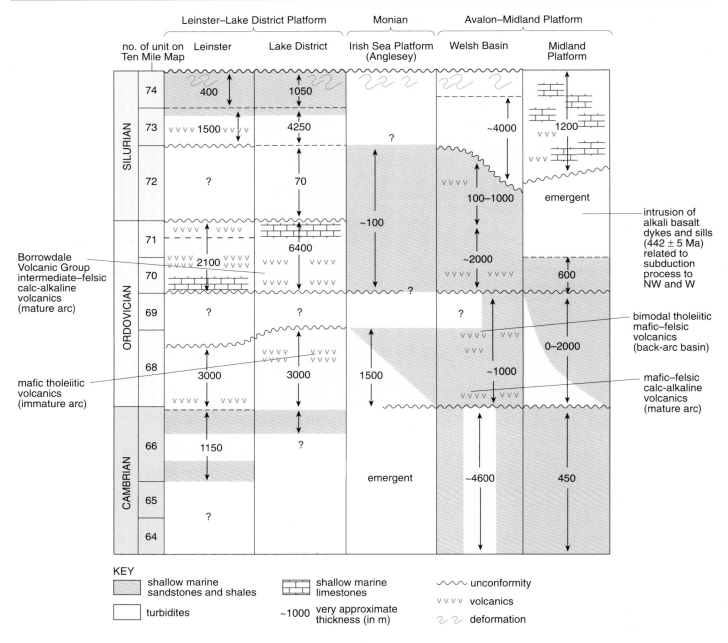

Figure 7.10 Summary of the Lower Palaeozoic lithological successions across the Avalon–Midland Platform, Monian and Leinster–Lakes Terranes. Note the vertical scale represents time and not unit thickness across the three terranes.

AVALON–MIDLAND PLATFORM TERRANE

Avalon–Midland Platform: During the Cambrian, a shallow-water marine succession dominated by mudstones (≤1 km) plus some minor sandstones and limestones formed across the Avalon–Midland Platform (Figure 7.10 and Plate 2). The succession is relatively uniform in thickness, with only minor variations occurring along a few active faults. In contrast, by the Early Ordovician, the rate and type of sedimentation across the terrane was controlled by rifting and associated subsidence, resulting in variations in sedimentary thickness of up to 2 km across some of the fault-bounded blocks (Figure 7.10 and Plate 3).

By the Mid-Ordovician rifting had ceased, returning the Avalon–Midland Platform to shallow-water marine conditions. From the Late Ordovician to the Mid-Silurian, the Midland Platform rose above sea-level, preventing sedimentation from occurring at this time. The intrusion of alkali basaltic dykes and sills (442 ± 3 Ma) in the north records a period of deep partial melting and the end of subduction, probably associated with the detachment of the subducting slab from the rest of the plate.

As igneous activity dwindled, the crust underwent **thermal relaxation** (because it was now less buoyant) and the Avalon–Midland Platform became submerged once more, forming a shallow-water marine shelf. This was accompanied by the deposition of an alternating sequence of mudstones and limestones (e.g. the Wenlock and Ludlow Series across Shropshire on the Welsh Border), interspersed with occasional volcanic ash horizons. The limestones formed in association with carbonate **build-ups** (up to 30 m high and 50 m across) that contained abundant corals and sponges, and were surrounded by shell and **crinoid** fragments. The stratigraphic record reveals that periodically, these carbonate build-ups (colloquially described as 'patch reefs') died back as a result of being smothered by mud and volcanic ash.

Welsh Basin: During the Early to Mid-Cambrian, crustal extension and the ensuing isostatic subsidence caused the Welsh Basin to change from a shallow-water marine environment to a deep oceanic slope that was dominated by turbiditic successions (Figure 7.10 and Plate 2). Conditions had stabilized by the Late Cambrian, allowing a shallow-water marine environment to return. By the Early Ordovician, mafic to felsic calc-alkaline volcanic successions, related to subduction and the development of a mature island arc along the northern margin of eastern Avalonia, superseded sedimentation in the south-east for a short period (~5–10 million years) (Figure 7.10 and Plate 3). By the Mid-Ordovician, the style of volcanism had changed and was now characterized by compositions ranging from tholeiitic basalts to rhyolites. This is typical of volcanism associated with extension that occurs behind a mature volcanic arc. These regions are referred to as **back-arc basins** and can be thought of as miniature sea-floor spreading environments (e.g. present-day Sea of Japan). Igneous activity in the back-arc basin continued into the Early Silurian until the subducting slab became fully detached. The northern margin of Avalonia now ceased to be an active continental margin, and deposition of turbidites down the steep continental slope resumed, forming strata up to 4000 m thick (Figure 7.10).

Monian Terrane

Irish Sea Platform: The absence of Cambrian sediments between south-east Ireland and Wales implies that at this time, this area formed an emergent landmass (Plate 2). The type of sedimentation was different on either side of this emergent platform, with deep marine conditions to the north and a shallow-water marine basin to the south. The platform grew laterally during the Ordovician and Silurian (Plates 2–3), with turbiditic successions around its outer margins in the Ordovician giving way to shallower-water marine sandstones and shales during the Silurian (Figure 7.10)

Leinster–Lakes Terrane

The Lake District and Isle of Man contain only minor Cambrian sediments, whereas Leinster is dominated by a thick Upper Cambrian to Middle Ordovician turbidite sequence and minor basaltic lavas (Figure 7.10). These formed on the slope leading down to the trench, which separated this terrane from the Iapetus Ocean to the north (Plates 2–3).

During the Ordovician and Silurian, thick turbiditic successions interspersed with volcanic materials and minor sandstones and limestones were deposited across most of this terrane. Although small volumes of basaltic tholeiitic lavas of the kind typical of an immature volcanic arc were erupted during the Cambrian, by the Early Ordovician these had been superseded by more voluminous calc-alkaline intermediate–felsic pyroclastic deposits (e.g. the Borrowdale Volcanic Group in the Lake District) indicative of a maturing volcanic arc and southward subduction under the Avalonian margin (Figure 7.10 and Plate 3). As volcanic activity waned and eventually ceased during the Silurian, thick turbidite deposits (including those at Tebay) were once more deposited down the steep continental slopes (Figure 7.10).

Therefore, throughout the Early Palaeozoic, the southern British Isles was dominated by the deposition of marine sediments in often rapidly subsiding basins, controlled by crustal extension. To maintain the high rates of sedimentation seen in the basins, the adjacent landmasses (e.g. the Midland and Irish Sea Platforms) must have undergone rapid uplift and subsequent erosion (Plate 2).

7.3.3 Volcanism and its effect on the tectonic evolution of the southern British Isles

Although the northern margin of Avalonia was initially passive during the Cambrian (Plate 2), by the Early Ordovician it had become an active continental margin, as the Iapetus Ocean was subducted southwards beneath it (Plate 3). Based on the different types of igneous activity described above (Figure 7.10), four major steps in the volcanic development of the southern British Isles during the Early Palaeozoic can be identified:

(1) *Early–Mid-Ordovician*: start of arc volcanism, initially in the Welsh Basin and subsequently in the Leinster–Lakes Terrane.

(2) *Mid-Ordovician*: after a brief hiatus in igneous activity, a back-arc basin developed rapidly in the Welsh Basin, associated with the northward migration of the volcanic front. Significant volumes of volcanic material were erupted, forming a series of volcanic islands. Over time, these became progressively surrounded by interbedded shallow-water marine sediments and **volcaniclastic** debris (e.g. the Snowdonia Volcanic Group, Plate 3).

(3) *Mid–Late Ordovician*: increasing volcanic activity and a shift from mafic lava flows to intermediate–felsic pyroclastic deposits occurred throughout the Leinster–Lakes Terrane (Figure 7.10). This change in composition and eruptive style can be attributed to the arc progressively maturing with time.

(4) *Late Ordovician*: volcanism (and subduction) ended in the Welsh Basin and Leinster–Lakes Terrane, with the detachment of the subducting slab from its plate (Figure 7.11c).

❑ Section 4.1 described how the volcanic front occurs directly above the point at which partial melting commences at a subduction zone (i.e. ~80–90 km depth, Figure 4.1b). If this depth-to-melting relationship is maintained but the angle of subduction is increased, what will happen to the position of the volcanic front?

■ It will migrate towards the trench (Figure 7.11).

In the case of the southern British Isles, the Iapetus Ocean was subducted southwards below the northern Avalonian margin. The northward migration of the volcanic front away from the Welsh Basin towards the Leinster–Lakes Terrane during the Ordovician is believed to be a result of just such an increase in the angle of subduction (Figure 7.11a–b).

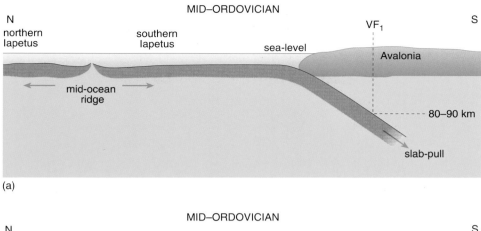
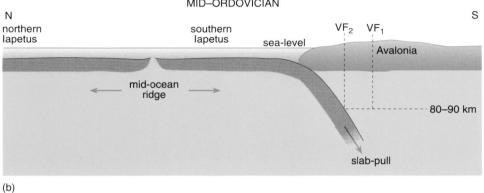
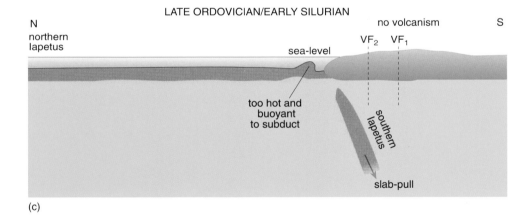

Figure 7.11 Cartoons showing the changing location of the volcanic front (VF) below the southern British Isles during the Mid–Late Ordovician, illustrating why the northern margin of Avalonia changed from a destructive to a passive margin.
(a) During the Mid-Ordovician, the rate of subduction below the northern edge of Avalonia was greater than the rate of new crust formation at the mid-oceanic ridge, resulting in the ridge moving towards the subduction zone.
(b) Over time, the angle of subduction became steeper. This resulted in the volcanic front moving towards the edge of the continental plate. (Migration is exaggerated in this cartoon)
(c) By Late Ordovician/Early Silurian times, the southern plate of the Iapetus Ocean had been completely subducted, causing the northern plate to collide with Avalonia. As the oceanic crust at this plate edge was still relatively hot and young, it was too buoyant to be subducted.

Faunal assemblages in the north and south of the Iapetus Ocean indicate that the ocean was at least 1000 km wide at the end of the Ordovician. This means that the cessation of volcanism at this time cannot be attributed to continental collision, so why did it stop? The currently accepted model is that the rapid rate of subduction below the active margin of Avalonia caused it to run over and consume the slowly spreading constructive plate boundary ('mid-oceanic ridge') of the Iapetus Ocean (Figure 7.11c). This is similar to what is currently happening in the east of the Pacific Ocean where constructive plate boundaries are being over-run by North America. Once the southern plate of the Iapetus had been subducted below Avalonia (primarily due to slab-pull processes), the edge of the northern plate, consisting of newly formed oceanic crust, was in contact with the continent. It was too hot and buoyant to be subducted. It may have accreted to the northern margin of Avalonia, uniting the two across a passive continental margin, or the junction may, for a time, have been a strike–slip fault (conservative plate boundary).

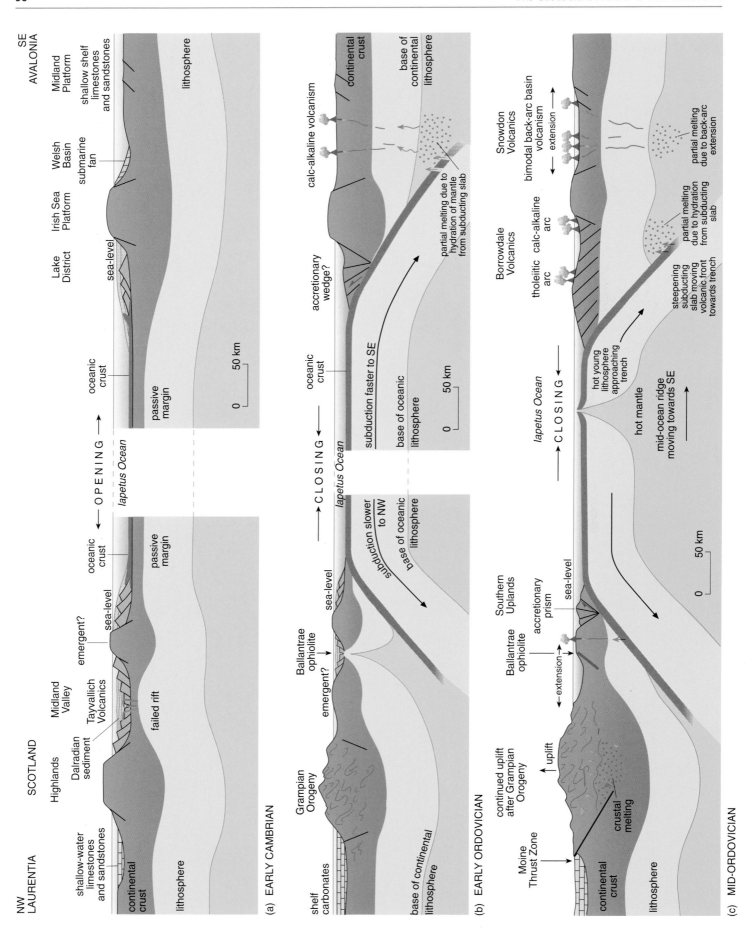

The Caledonian Orogenic Belt

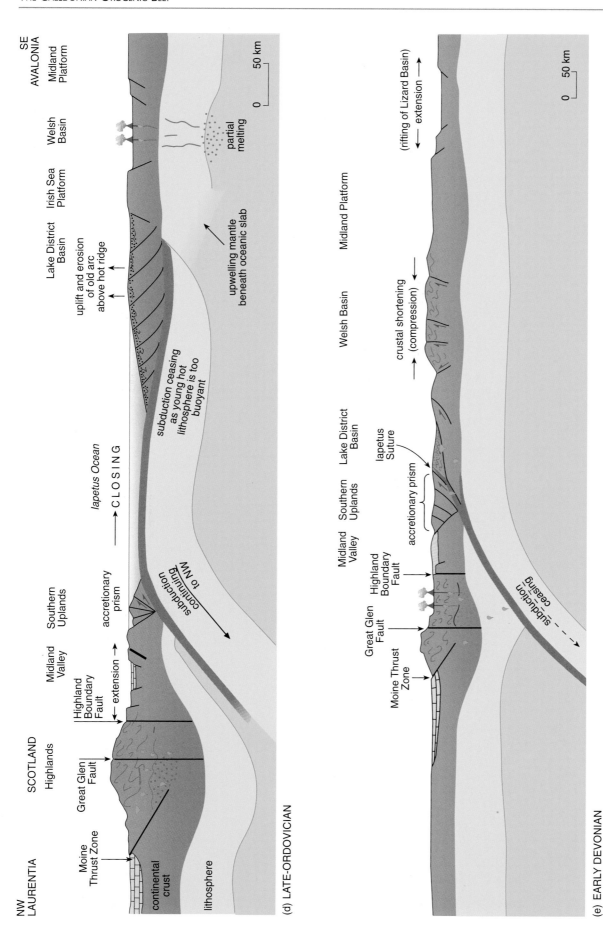

Figure 7.12 Model of the progressive migration of the northern and southern British Isles as the Iapetus Ocean first opened and then closed from the Cambrian to Early Devonian, ultimately resulting in the development of the Caledonides. (Note change of scale in (e).)

7.4 The site of the Iapetus Suture

With the cessation of subduction along the southern margin of the Iapetus (but continued subduction on the northern side, where the Southern Uplands accretionary prism was still being stacked up, see Figure 7.8), it was only a matter of time before Avalonia (and the southern British Isles) collided with Laurentia (and the northern British Isles).

Given the evidence for the existence of subduction of the Iapetus northwards beneath the Southern Uplands (Section 7.2.4), and southwards beneath the Leinster–Lakes Terrane and Welsh Borders (Section 7.3.3), the suture zone, uniting the two formerly separate continental masses, must occur somewhere in the Solway Firth area. The only exposure of the suture is in eastern Ireland. It is buried elsewhere by Carboniferous and younger sediments. However, the general location of the suture has been confirmed by studies of **trilobites** and other fossil fauna, which show discrete and distinct species occurring in rocks on either side of the suture up until the time when the ocean had become sufficiently narrow for species to migrate and co-evolve across the whole basin. In addition, deep seismic surveys in the Irish Sea and North Sea reveal a buried thrust surface dipping north-westward at about 20°, showing where the leading edge of the Avalonian continental crust was dragged below the Southern Uplands Terrane in the final stages of collision.

Collision between the two continents occurred during the end of the Silurian to the beginning of the Devonian, causing regional deformation and low-grade metamorphism of the terranes on either side. Although the extent of folding varied from tight isoclinal folds in the Southern Uplands to open, gentle folds across the Welsh Basin and Avalon–Midland Platform, metamorphism was relatively constant across all terranes, converting the shales and greywackes into low grade **slates**.

A possible model for the lithotectonic evolution of the northern and southern British Isles throughout the Early Palaeozoic culminating in how the Iapetus Ocean closed is shown in Figure 7.12.

7.5 Post-orogenic Caledonian granites

The terranes affected by the Caledonian Orogeny are intruded by granite plutons, many of which display a cross-cutting relationship with the Caledonian structures. This cross-cutting relationship is particularly apparent in the Southern Uplands.

> Question 7.7 From these relationships, what can you deduce about the age relation between the granites and the Caledonian structures?

As these granites do not exhibit any signs of the regional deformation or metamorphism characteristic of the Caledonian units, they can be described as post-orogenic, i.e. occurring after the orogeny. These post-orogenic granites are not just confined to the area north of the suture zone but can also be seen in the Lake District.

Geophysical and borehole evidence has shown that similar granites extend beneath the Carboniferous cover of the Pennines. Radiometric dating suggests that they formed around 400 Ma (Early Devonian). The Shap Granite at the edge of the Lake District (NY(35)5509, Figure 7.13) forms the most easterly outcropping post-orogenic granite.

> Question 7.8 With reference to Section 4.2.2, briefly describe how these post-orogenic granites could have formed.

Figure 7.13 Shap Granite, Cumbria, one of several granites that link down to a large batholith, underlying the north-west of England.

7.6 Summary

Northern British Isles (including the Southern Uplands)

- Throughout the Precambrian to Early Palaeozoic, the northern British Isles was located on the southern margin of Laurentia, slowly drifting northwards from ~40° S to ~20° S.
- The Moine Supergroup (1005–873 Ma), which is predominantly found in the Northern Highlands Terrane, consists of a repeating succession of shallow-water marine to marginal shelf deposits. These deposits experienced two periods of regional metamorphism and deformation at ~850–800 Ma and 475–463 Ma, associated with crustal shortening and thickening, and arc accretion.
- The Dalradian Supergroup (<850–509 Ma) currently found in the Central Highlands Terrane, consists of shallow-water marine to marginal shelf sediments, interbedded with the Tayvallich Series (MORB-like basaltic sills and dykes). These volcanics and intrusives are associated with extensive crustal stretching and rifting, which led to the eventual development of sea-floor spreading further south and the formation of the Iapetus Ocean during the Late Proterozoic (~595 Ma).
- Throughout the Ordovician and Silurian, cherts, shales and greywackes were deposited in the ocean basin and adjacent trench, with subduction and convergence thrusting these deposits up into an accretionary prism (now represented by the Southern Uplands Terrane). At Ballantrae, a section of oceanic crust and upper mantle has been thrust onto the continent, forming an ophiolite.
- The Grampian Orogeny (467–455 Ma) was associated with the convergence of a small volcanic arc with Laurentia, causing extensive NW–SE compression across the region. This collision resulted in a switch in tectonic process from sea-floor spreading to subduction, leading to the eventual closure of the Iapetus Ocean.

Southern British Isles

- Between the Precambrian to Late Ordovician, the southern British Isles drifted northwards as part of the Avalonian microcontinent, from ~60° S to ~30° S.
- Up until the Early Ordovician, the northern margin of Avalonia was volcanically active, characterized by tholeiitic basaltic lava flows, typical of an immature arc. This activity started in the Welsh Basin before moving to the Leinster–Lakes Terrane and was short-lived, lasting for only 5–10 million years.
- After a brief hiatus during which time the volcanic front moved northwards, the Welsh Basin switched to become a back-arc basin characterized by tholeiitic basaltic to rhyolitic volcanism associated with crustal extension. In the Leinster–Lakes Terrane, the arc continued to mature, erupting calc-alkaline intermediate–felsic pyroclastic material.
- By the Late Ordovician, subduction and the associated volcanism stopped because the constructive plate boundary was over-ridden by the northern margin of Avalonia, which became essentially a passive continental margin. (Meanwhile, northward subduction continued below Laurentia.)
- Throughout the Late Ordovician to Silurian, the type and rate of sedimentation that occurred across the southern British Isles was controlled by episodic periods of crustal extension and rifting, with sediments deposited in shallow to deep-water marine basins.
- Collision between Laurentia and Avalonia occurred during the Late Silurian to Early Devonian, causing crustal thickening, regional folding and metamorphism followed by the emplacement of post-orogenic granites.

8 THE OLDER COVER

8.1 INTRODUCTION

The Devonian and Carboniferous of the British Isles can be divided into two lithotectonic units (see Appendix):

- *The Older Cover* consisting of relatively undeformed sediments that overlie the Caledonian Orogenic Belt. (You should note that on a local scale, some deformation might be present in these units.)
- *The Variscan Orogenic Belt* which consists of Upper Palaeozoic sediments and some minor volcanics that have been deformed by the Variscan Orogeny. In some literature the Variscan is also referred to as the 'Armorican' or 'Hercynian' orogeny. In the British Isles, the effects of this orogeny extended from the Late Devonian to Late Carboniferous, with the most intense deformation limited to the south-west of England, south Wales and south-west Ireland.

❑ Looking back at Section 1 why do we sometimes refer to a part of a Period as 'Upper' and sometimes 'Late'?

■ The term 'Upper' refers to the succession of rocks. 'Late' refers to a period of time. Thus Upper Palaeozoic rocks were formed during the Late Palaeozoic.

The Older Cover is the main topic of discussion in this Section, while the effects and cause of the Variscan Orogeny are examined in Section 9. Although the Permian Period belongs to the Palaeozoic Era (see Figure 2.1), as it is part of the Younger Cover it will be discussed in Section 10.

Table 8.1 summarizes the distribution of Devonian and Carboniferous sedimentary and volcanic successions in eleven specified areas across the British Isles. The ticks indicate:

(a) which Carboniferous units are present in each area;

(b) which Devonian/Old Red Sandstone units are present in each area;

(c) whether Devonian or Carboniferous volcanic activity is present in each area.

Table 8.1 overleaf, which illustrates the development of the Older Cover across the British Isles, emphasizes differences in the sedimentary environment of deposition during the Devonian. In south-west England, all sediments referred to as *Devonian* were deposited under predominantly marine conditions, whereas *Old Red Sandstone* deposits of similar age found north of Devon were all deposited in continental or freshwater environments on land.

Question 8.1   Using Table 8.1, where are outcrops of the Middle Old Red Sandstone/Devonian found in the British Isles?

The region in northern Scotland where the Middle ORS is found is known as the Orcadian Basin. During the Devonian, this was a shallow-water **ephemeral** fluvial and lake environment, within an otherwise arid landscape and flanked by mountains created by the Caledonian Orogeny (Figure 8.1, overleaf). In south-west England, the Middle Devonian consists of a marine succession deposited in an area that underwent fluctuations in the water depth (Plate 4).

Question 8.2   What could the lack of Middle Devonian outcrop in the rest of the British Isles represent?

Table 8.1 Summary of Devonian and Carboniferous rock unit distribution in the British Isles.

1. Moray Firth NH(28)
2. Midland Valley NS(26)–NO(37)
3. South-west Southern Uplands, Dumfries and Galloway NX(25)–NZ(45)
4. Northumberland NY(35)–NZ(45) and north-east Southern Uplands NT(36)
5. Lake District and Cross Fell NY(35)
6. North Pennines, e.g. Ingleton (SD(34)6973)
7. North Wales, Denbigh (SJ(33)0567)– Oswestry (SJ(33)2929)– Wrexham (SJ(33)3550)
8. Nuneaton (SP(42)39)
9. Dyfed, Pembs, SM(12)–SN(22)), south Wales coalfield SS(21)–ST(31)
10. Mendips (ST(31)46–64)
11. South-west England SW(10)–ST(21)

	1	2	3	4	5	6	7	8	9	10	11
(a) CARBONIFEROUS											
Westphalian and Stephanian	–	–	–	–	✓	✓	✓	✓	✓	–	✓
Westphalian (Coal Measures)	–	✓	✓	✓	✓	✓	✓	✓	✓	✓	✓
Namurian (Millstone Grit)	–	✓	–	✓	✓	✓	✓	–	✓	–	✓
Tournaisian and Viséan (Carboniferous Limestone Series)	–	✓	✓	✓	✓	✓	✓	–	✓	✓	✓
(b) DEVONIAN/ORS											
Upper ORS	✓	✓	✓	✓	–	–	–	–	✓	✓	✓
Devonian Limestone	–	–	–	–	–	–	–	–	–	–	✓
Middle ORS/Devonian	✓	–	–	–	–	–	–	–	–	–	✓
Lower ORS/Devonian	✓	✓	–	✓	–	–	–	–	✓	–	✓
(c) VOLCANIC ACTIVITY											
Carboniferous volcanics	–	✓	–	✓	–	–	–	–	–	–	✓
Devonian volcanics	–	✓	✓	✓	✓	–	–	–	–	–	✓

Note: In south-west England (11), there is a complete Devonian succession consisting of marine and Old Red Sandstone strata.

By referring back to Table 8.1, it is apparent that the entire continental ORS is missing across most of the Southern Uplands, all of the Lake District and the northern Pennines (exemplified by Ingleton). These regions are believed to have been emergent throughout the Devonian (Plates 4 and 5). Any ORS sediment that was deposited must have been subsequently eroded and removed from the area (probably as a result of uplift), before the Carboniferous sediments were deposited on top. The whole of the ORS is also absent across most of north Wales, excluding Anglesey. Detailed field studies have shown that during the

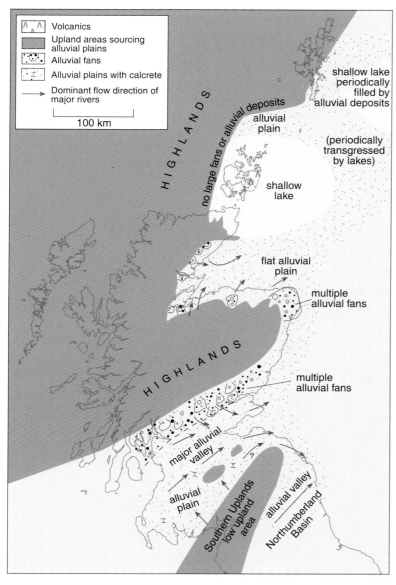

Figure 8.1   Palaeogeography of Scotland during the Mid–Late Devonian, indicating the main areas of sub-aerial deposition, sourced from adjacent highlands. *Note*: the buff-coloured area denotes lowland. (From Craig, G.Y. (ed.) (1991) *Geology of Scotland*, Geological Society.)

Devonian, this part of the Caledonian Orogenic Belt was still being uplifted (related in part to the emplacement of the Leinster Granite), forming a rising land area and preventing sedimentation from taking place (Plate 4).

By the Late Devonian (Plate 5 and Table 8.1), sedimentation had recommenced across most of the British Isles. Continental ORS deposits of this age are found across Scotland, the north of England and north-east Ireland. The environment of deposition was marine across the south of England, southern Wales and south-west Ireland. There is however a broad belt across the Midlands of England, most of Wales and eastern Ireland where no Upper Devonian sediments are found. A similar area devoid of Upper Devonian sediments occurs in the Galway to Donegal region of western Ireland (Plate 5). The lack of sediments in these two areas can be explained by these regions forming highland areas during the Late Devonian.

8.2 Interpreting the Devonian sedimentary environments

8.2.1 Old Red Sandstone continental deposits

Question 8.3 Figure 8.2 shows four logs of ORS successions from Dyfed (south Wales) and the Midland Valley of Scotland. Briefly describe any pattern you can see in the grain size variations in the logs.

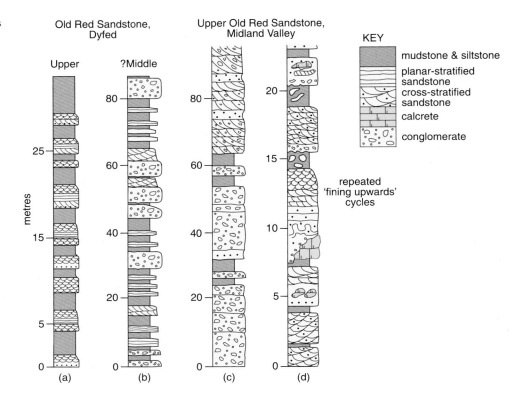

Figure 8.2 Generalized **graphic logs** of Old Red Sandstone successions for Dyfed and the Midland Valley of Scotland. The width of the columns is roughly proportional to the grain size, although the vertical scale of each column is not identical.

In column (a) and to a lesser extent (d), the fining-up trend in grain size is accompanied by a change in the type of sedimentary structures that are present. These vary from large-scale cross-stratification to **planar stratification** (Figure 8.3a) and small-scale cross-lamination. These features are a clear indication of an upward decrease in current velocity (i.e. a decrease in current velocity with time), culminating in the deposition of silts and suspended muds. This lithological and sedimentary structural assemblage is characteristic of a fluvial environment of deposition.

The conglomerates in columns (b) and (c) (Figure 8.3b) represent high-energy conditions, and were deposited by flash floods and storms of the kind that rapidly transport and dump the unsorted material. These deposits then fine up into a succession of cross-stratified and planar-stratified sands, which are interbedded with silts. This succession is typical of an alluvial fan deposit that forms at the edge of mountainous regions.

(a) (b)

Examples of these alluvial, fluvial and lake deposits can be found in the ORS strata anywhere north of Devon in the British Isles. In general, the Lower and Upper ORS in the Welsh Borderlands and Midland Platform are characterized by high-energy flash flood conglomerates that fine up into fluvial sandstones, siltstones and mudstones. In the Midland Valley, numerous Lower and Upper ORS alluvial fan deposits occur along its northern and southern margins, with material sourced from the edge of the Scottish Highlands and Southern Uplands mountains respectively (Figure 8.1). This contrasts with the centre of the valley, which is characterized by a fluvial floodplain. Here Upper ORS sedimentary successions are up to 3 km thick in Scotland, increasing to 6 km between Limerick and Waterford in Ireland. The thickness of these deposits was controlled by two main factors – contemporaneous strike–slip faulting along the Highland Boundary Fault increasing the depth of the sedimentary basin, and isostatic subsidence of the crust in response to rapid deposition of sediment from the highlands into the basins.

Way to the north of the Midland Valley, the Orcadian Basin (Figure 8.1 and Plates 4–5) exhibits a progressive shift from ephemeral lake deposits in the Lower ORS to thicker lake deposits (rich in fossil fish) associated with fluvial deposits in the Upper ORS. These are fringed by aeolian (wind-deposited) dunes, which grade westwards into conglomerates and breccia deposits, as the mountainous source regions are approached.

Figure 8.3 (a) Centimetre-scale cross and planar stratification in ORS sandstones, Yesnaby, Stromness, Orkney. The sequence fines up from medium cross-stratified sandstone into a finer planar-bedded sandstone, ending with a flaggy, unbedded siltstone (at very top of photograph). (b) Unsorted ORS conglomerate, Downie Point, Stonehaven, Aberdeenshire, resting on a medium–fine-grained micaceous sandstone. The conglomerate is predominantly composed of boulder-sized clasts of quartzite, plus smaller clasts of igneous rocks such as granite, rhyolite and andesite. (For scale, the large round clast on the right hand side of the photograph is ~85 cm long.)

> Question 8.4 As their name suggests, unlike normal fluvial to lake sediments that are yellow to grey, the Old Red Sandstone strata found north of Devon are generally red–brown in colour. What is likely to be the cause of this coloration?

The red coloration of the ORS is not the only indication that an arid environment was prevalent at the time of deposition. At the top of some successions (especially in the Midland Valley), carbonate evaporite deposits (known as **calcrete** or **caliche**) are found. These represent fossil soil horizons that formed under arid conditions. They consist of fine-grained nodular to unbedded limestones that have precipitated around and between siliciclastic grains in the soil by evaporation processes.

Palaeomagnetic studies have shown that during the Devonian, the British Isles was at a latitude of ~15°–20° S. This is just north of the present-day deserts in South Africa (e.g. the Kalahari and Namibian deserts) and Western Australia (e.g. Gibson desert).

8.2.2 Devonian marine deposits

In contrast to the arid conditions that prevailed across most of the British Isles during the Devonian, the region south of north Devon was characterized by a fluctuating marine environment, situated at the edge of the Rheic Ocean (Figure 3.1c–d).

North Devon marks the fluctuating boundary between continental and marine depositional environments during the Devonian, with 3.5–5.5 km of alternating continental ORS and marine Devonian strata in this region. Throughout the Early to Late Devonian, the marine successions in north Devon are represented by fine-grained laminated shales and sandstones and contain marine fossils such as **brachiopods**, **bivalves**, **echinoids** and trilobites (Plates 4–5). These units grade into cross-stratified siltstones and sandstones, which are often draped by fine mantles of mud. Although these sandstones and siltstones are devoid of the shelly fossils found in the strata below, they do contain abundant **trace fossils**, plant and vertebrate remains.

This combination of strata, sedimentary structures and fossil types represents a periodic decrease in water depth leading to a switch from a marine to a continental environment, equivalent to an estuary or delta. By the end of the Late Devonian, the depositional environment in north Devon had reverted back to predominantly marine conditions.

In south Devon, the Early to Late Devonian marine successions represent a shallow-water sequence of sandstones, siltstones, mudstones and limestones, interspersed with variable thicknesses of volcanic ash (Plates 4–5). Throughout the Early Devonian, the fossiliferous limestone beds were relatively thin, and can be seen to be **carbonate platform** deposits. By Mid-Devonian times, although some of these thin carbonate bank deposits were still forming, they were joined by much more massive limestone deposits, the shape and fossil content of which are indicative of reef-like carbonate build-ups. In the Late Devonian, carbonate deposition gave way to the formation of green and grey mudstones (subsequently metamorphosed to slate), interbedded with thin nodular limestone beds.

Further south-west in Cornwall, where the sedimentary basin was deeper throughout the Devonian, the stratigraphic succession is dominated by black shales grading into mudstones, siltstones and sandstones in north Cornwall, and greywackes interbedded with thin limestones in south Cornwall (Plate 5). There are also some small outcrops of pillow basalts and pyroclastic deposits in Cornwall. These are geochemically similar to MORB and indicate proximity to the Rheic Ocean, although their tectonic relationships are both complex and unclear. The more obvious ocean floor rocks in the thrust complex of the Lizard ophiolite are discussed in Section 9.2.3.

8.3 Development of the Devonian highlands

The previous Section described how the distribution and thickness of Devonian sediments across the British Isles was strongly controlled by the occurrence of extensive highland regions. These developed across the Caledonian Orogenic Belt because the collision of Avalonia and Laurentia caused crustal thickening, which led to rapid isostatic uplift and the emplacement of hot, buoyant granitic plutons (e.g. the Shap Granite, Figure 7.13). In the Early Devonian, an extensive highland region across the Southern Uplands, the north of England, and the Galway to Donegal region of western Ireland (Plate 4), had a relief comparable to the present day Himalayas. This mountain belt continued to rise and spread laterally throughout the Mid and Late Devonian (Plate 5), forming an extensive highland area across the Midlands of England, Wales and eastern Ireland, along with another major highland over western Ireland, referred to as the Galway–Mayo–Donegal High.

As these mountains rose, they underwent rapid erosion, with much of the eroded material deposited in the adjacent sedimentary basins, which caused them to subside. As the tops of the mountains were worn away, **isostatic compensation** resulted in the base of the crust below the mountain belt rising. This combination of erosion and uplift of the Caledonian mountains and filling in of adjacent basins by sediment led to a general levelling out of the British Isles topography, leaving a few low-relief upland areas.

8.4 Igneous activity in the Devonian

> Question 8.5 Using Plates 4 and 5, which regions of the British Isles were affected by Devonian volcanic and plutonic activity?

Although some of the granites on either side of the Iapetus Suture Zone are contemporaneous with continental collision and the Caledonian Orogeny, the volcanic activity and many other intrusive rocks found in the Southern Uplands to central Highlands relate to neither the subduction nor the collision processes associated with the closure of the Iapetus Ocean. This activity actually relates to localized extension and strike–slip movement along major faults (e.g. the Highland Boundary and Great Glen Faults). This resulted in the formation of small **pull-apart basins**, and also allowed melts that formed within and below the newly sutured continental plate to be channelled upwards through the crust. Arrival of magma at shallow levels resulted in **caldera** subsidence and the intrusion of **ring dykes**, which were associated with both **effusive** and **explosive** volcanism. Examples include the ring complexes of Glen Coe (NN(27)15) and Ben Nevis (NN(27)17), and the volcanic and intrusive complex in the Cheviot Hills (NT(36)91).

At several locations within the Midland Valley, the Devonian volcanic successions are found interdigitated with ORS sediments (e.g. the Ochil and Sidlaw Hills in Fife and Angus, Figure 8.4a), with eroded and rounded igneous **clasts** frequently found in the coarser sedimentary units.

> Question 8.6 What does the interdigitation of ORS sediments and igneous clasts imply about the rate of erosion at this time?

Therefore during the Devonian Period, the British Isles was characterized by the formation and uplift of an extensive mountainous region undergoing rapid erosion, the products of which were deposited in adjacent rapidly subsiding basins. The highland areas formed as a result of continental collision and rose isostatically as the crust became thickened by collision, with this enhanced by the emplacement of post-orogenic granites. These upland areas can be referred to as blocks, with the intervening sedimentary basins called troughs.

Figure 8.4 (a) Distribution of Devonian and Carboniferous volcanic and shallow-level intrusions in the Midland Valley.
(b) The Rock and Spindle near St Andrews, Fife. This shallow-level intrusive feature acted as a feeder system, supplying magma to local small vents.
(c) Aerial view of Salisbury Crags, Edinburgh, a major Early Carboniferous volcanic centre which fed smaller vents including the Castle Rock. (From Francis, E. H. (1991) 'Carboniferous–Permian igneous rocks', in G. Y. Craig (ed.) *Geology of Scotland*, Geological Society.)
(d) Layered pyroclastic deposits at Elie Harbour, East Fife, forming part of a small volcanic cone.

8.5 Igneous activity in the Early Carboniferous

During the Early Carboniferous, less extensive and very localized mafic to intermediate volcanic activity occurred in Dumfries and Galloway, the Lake District, Derbyshire and the Mendips (Plate 6). Products of these eruptions are frequently interbedded with contemporaneous sedimentary deposits. The major sites of volcanic and shallow intrusive activity at this time were in the Midland Valley (e.g. the Clyde Plateau Lavas, the East Fife Volcanics and Lothian Volcanics, Figure 8.4a). The majority of these volcanics were erupted from a swarm of small (<500 m) ephemeral **vents** (many of which can still be seen, e.g. Figure 8.4b), forming localized overlapping deposits (Figure 8.4d). A few larger central volcanoes also existed (e.g. the Lomond Hills, Salisbury Crags and Edinburgh Castle, Figure 8.4c), which acted as the main reservoirs for many smaller cones. Geochemical and other studies indicate that this region formed a passive continental rifting environment that was driven by lithospheric extension (see Section 4.2.1 and Figure 4.2a).

8.6 The Carboniferous transgression

Question 8.7   Refer to Table 8.1, then describe in one sentence how the distribution of the Carboniferous strata differs to that of the Devonian strata.

During the Late Devonian to Early Carboniferous, a rise in global sea-level (Figure 3.2) flooded the continental ORS along with many of the previously emergent tracts of Lower Palaeozoic rocks. As a result, these areas became overlain by the marine strata of the Carboniferous Limestone Series (80). The Carboniferous Limestone Series is not just limestone. It consists of a variety of rock types (Figure 8.5), the origin of which will be discussed in Section 8.6.2. From now on, we will refer to the Carboniferous Limestone Series as the Lower Carboniferous*. Let us now consider the age of its base across the British Isles.

Figure 8.5   Lower Carboniferous (Dinantian) logs illustrating differences in thickness, facies and age between the blocks and troughs of northern England. (Levels 1–6 on each log represent time equivalent units in the Dinantian.)

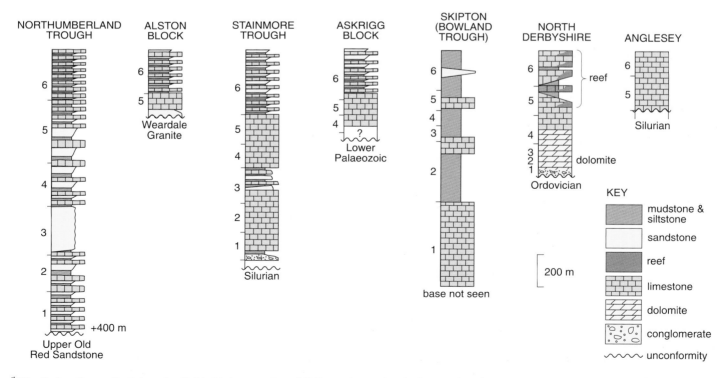

* The Carboniferous Period can be divided into a number of different time units. At the simplest level, the Lower Carboniferous is roughly equivalent to the Dinantian Series. The rest of the Carboniferous Period forms the Upper Carboniferous or Silesian Series, which comprises the Namurian Stage and Westphalian Stage. Alternatively, the Carboniferous Period can be referred to as the Lower Carboniferous (Dinantian), Middle Carboniferous (Namurian) and Upper Carboniferous (Westphalian).

Question 8.8 Establish the relative ages of the *oldest* Lower Carboniferous strata at each locality on Figure 8.5. From this, decide whether the Early Carboniferous transgression was abrupt or gradual. (The localities are arranged in a north–south order, but for this exercise you do not need to know their precise locations. The numbers at the side of the columns are the biostratigraphical divisions of the successions (i.e. they refer to approximate 'time slices'), which can be used to correlate successions between columns.)

The age of the base of the Lower Carboniferous across the British Isles is dependent on the topography of the land at this time. This consisted of a series of upstanding blocks, which remained above sea-level for much of the Early Carboniferous. These blocks were separated by low-lying troughs, which were flooded earlier by the rising Carboniferous seas (Figure 8.6 and Plate 6). By the Late Dinantian–Early Namurian (Early–Mid-Carboniferous), high rates of erosion resulted in only a few of the isolated upland areas remaining above sea-level.

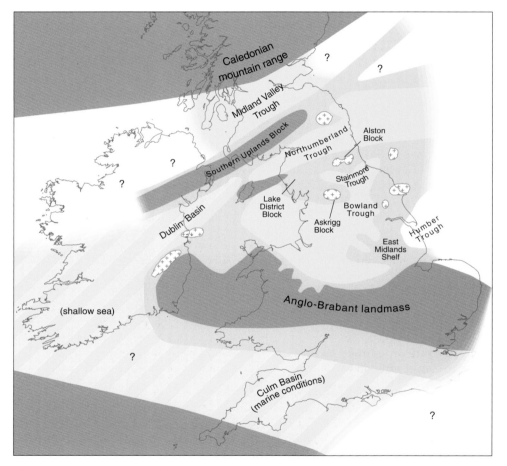

Figure 8.6 The British Isles during the Early Carboniferous (~345 Ma), indicating the palaeogeographic distribution of persistent highland areas, smaller buoyant blocks and intervening low-lying troughs. All the Carboniferous rocks, including the blocks and shelves, were submerged to different depths over time.

Figure 8.7 is a north–south transect between the Midland Valley and a tract of land known as the Anglo–Brabant landmass (formerly known as St. George's Land) which extends across the Midland Platform and Welsh Basin, illustrating how the thickness of Carboniferous strata varies between the troughs and blocks. In general, the troughs can be seen to contain relatively thick (and complete) Lower Carboniferous successions, whereas over the blocks, the Lower Carboniferous is relatively thin (and incomplete). The Middle Carboniferous (Namurian Series) thickness is relatively uniform across the blocks and troughs in the north of England (Figure 8.7), although it is much thicker in the Bowland Trough between the Askrigg Block and the Anglo–Brabant landmass. From the Namurian to the start of the Westphalian (Late Carboniferous), the only blocks that remained above sea-level were the Southern Uplands Block and the Anglo–Brabant landmass (Figure 8.7, Plate 7).

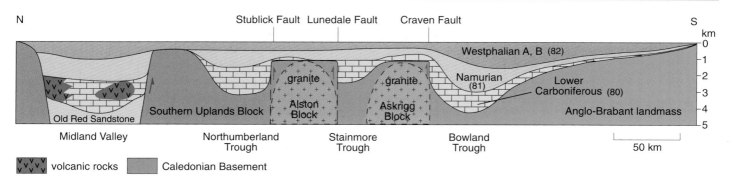

Figure 8.7   North–south cross-section from the Midland Valley to the Anglo–Brabant landmass, illustrating the block and trough structure and variable sedimentary thicknesses of Carboniferous strata above the Lower Palaeozoic basement.

8.6.1 Extensional tectonics

The start of the Carboniferous Period was marked by a series of global (eustatic) sea-level rises, which were associated with fluctuations in the size of ice sheets on Gondwana (still located over the southern pole). At the same time, the British Isles experienced significant climatic changes, going from arid to semi-arid and then monsoonal conditions (Figure 3.2), as the region drifted northwards across the equator and global temperatures increased (Figure 3.1d–e). These eustatic sea-level rises increased the potential number of submerged locations available for sedimentary deposition, while the changing climate varied the rate of erosion of adjacent continental landmasses (Figure 3.2).

> Question 8.9   Refer to Figures 8.6–8.7 and Plate 4. What common features underlie many of the blocks across the Southern Uplands and north of England?

These are post-orogenic granites and were an extremely important factor in controlling the locations and behaviour of the blocks, because their emplacement produced an overall decrease in local crustal density. In response to deposition of sedimentary material in the troughs, the increased weight caused those areas to subside in order to regain their isostatic equilibrium. On the other hand, the less dense granites made the blocks relatively buoyant so that when eventually sediment did begin to be deposited over the blocks, the rate of subsidence, and hence the rate of sediment accumulation, was slower. During the Early Carboniferous, erosion from the blocks also encouraged their isostatic rise.

The relative vertical movements of the blocks and troughs eventually resulted in faulting, with large-scale E–W trending faults forming along the margins of the blocks. The Alston and Askrigg Blocks are bounded by faults that were active from the Early Carboniferous, with the Craven Fault along the southern margin of the Askrigg Block exerting a significant control on the thickness of Mid–Upper Carboniferous sediments in the adjacent Bowland Trough (Figure 8.7).

To the south of the Pennine blocks, although no major granitic emplacement occurred beneath the broad Anglo–Brabant landmass (Figures 8.6 and 8.7), the crust here was still sufficiently thickened by the Caledonian Orogeny to maintain an overall buoyant signature. The Anglo–Brabant landmass continued to form an emergent continental block throughout the Carboniferous, influencing subsequent tectonic events associated with the Variscan Orogeny and closure of the Rheic Ocean to the south.

At the beginning of the Carboniferous, in addition to the isostatic readjustment associated with the erosion of the highland areas and deposition of materials in the adjacent basins, another important factor was at play, namely, the crust was beginning to recover from the effects of crustal shortening and uplift produced

(a) Devonian–Early Carboniferous

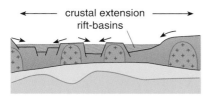

(b) Early Carboniferous

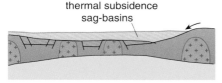

(c) Mid–Late Carboniferous

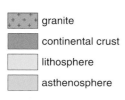

- granite
- continental crust
- lithosphere
- asthenosphere

Figure 8.8 Development of half-graben basins formed under an extensional tectonic regime, into thermal sag-basins during a period of crustal relaxation and subsidence.
(a) Devonian–Early Carboniferous. Crustal shortening and uplift produce a series of upland regions, some of which are underlain by granitic plutons.
(b) Early Carboniferous. Following on from the Caledonian Orogeny, northern England and the Midland Platform undergo a period of crustal extension due to thermal relaxation and isostatic readjustment, with troughs and rift-basins opening on either side of the upland blocks. These troughs continue to deepen predominantly due to rifting, while being filled with sediments.
(c) Mid–Late Carboniferous. Crustal extension ceases, ending rifting and slowing down the rate of subsidence in the troughs. Sediment continues to pour into the troughs, which continue to deepen but at a slower rate due to thermal relaxation, forming sag-basins. (From Leeder, M. R. and McMahon, A. H. (1998) in *Geology of England and Wales*, Geological Society.

by the Caledonian Orogeny (Figure 8.8a). As this thickened crust cooled, it started to subside and underwent a period of lithospheric extension. This extension possibly occurred as a result of the area's back-arc setting relative to the northward subduction of the Rheic Ocean, way to the south below central France and southern Germany. In any case, it contributed to the formation of a series of subsiding basins on either side of the buoyant blocks. These basins took the form of extensional grabens* and half-grabens* as faults formed along the margins of the blocks (Figure 8.8b). Throughout the Early Carboniferous, troughs continued to deepen because of extension and continued isostatic readjustment, while simultaneously being filled by sediments sourced from the adjacent upland blocks.

By the Mid to Late Carboniferous, crustal extension ceased, while subsidence in the troughs continued in response to sedimentation. The rate of sedimentation increased because sediment was now being supplied from both the rejuvenated and uplifted highlands in the north and the new highland areas in the south, as the Rheic Ocean closed and the Variscan Orogenic Belt moved northwards. The added weight of this material into the troughs, along with continued thermal relaxation of the crust, resulted in the formation of **sag-basins** across the region (Figure 8.8c). Therefore, during the Carboniferous, crustal movements associated with lithospheric extension as well as thermal relaxation resulted in local (epeirogenic) sea-level changes, controlling the location and rate of sedimentation.

To summarize, the Carboniferous highlands (Southern Uplands, Anglo–Brabant and Welsh landmasses) were created as a result of crustal shortening and uplift during the Caledonian Orogeny. These highland areas persisted while crustal extension during the Early Carboniferous created a series of rift-basins around them. Over time, these basins were progressively filled. During the Mid to Late Carboniferous, although the contribution to basin subsidence due to rifting decreased, a slower rate of thermal subsidence took over, which accompanied by increased sedimentation rates resulted in the formation of sag-basins across the region. The observed variations in Carboniferous sedimentation can therefore be attributed to the progressive development of a series of extensional rift-grabens, half-grabens and troughs across northern and central England on a local scale, as well as to global eustatic sea-level oscillations associated with variations in the size of the south polar ice sheet on Gondwana.

8.6.2 Carboniferous deltas

On Figure 8.5, the Lower Carboniferous successions can be seen to consist of an upward repetition of limestone, shale and sandstone. This repeating succession, or cycle as it is known, dominates the whole of the Lower Carboniferous in the Northumberland and Stainmore Troughs, and is also present on the top of the Alston and Askrigg Blocks (Figure 8.5–8.7).

A typical (and complete) cycle begins with marine limestone containing abundant crinoids, corals and brachiopods. The presence of limestone indicates that the depositional environment was characterized by warm, clear, shallow waters with only low rates of detrital sedimentation. As can be see on Figure 8.9, such limestone is then superseded by deposition of a coarsening-up siliciclastic succession of mudstones, siltstones and planar and cross-stratified sandstones. This represents an increase in the energy of deposition and also a shift from marine to non-marine conditions, which are increasingly dominated by the input of alluvial material. Sedimentary structures found within the sandstones indicate that they were fluvially derived, and were deposited in a series of

---

* Graben is the term given to fault-bounded troughs; a half-graben is faulted on one side. The term 'trough' is a generic term that can be used to describe non-faulted basins, or basins whose margin-type is unknown.

channels. As the river flowed through these channels, different scales of ripples formed, now preserved as variably cross-stratified beds. In the top part of the cycle, the depositional conditions change once more with the formation of mudstones sometimes topped by a thin coal seam. This represents a dramatic decrease in the energy of deposition with the mudstones and coal being formed by the accumulation of vegetation in a swamp or marsh. Each cycle is ended abruptly by subsidence or a eustatic sea-level rise, flooding the delta and returning to more marine conditions. The cycle then repeats.

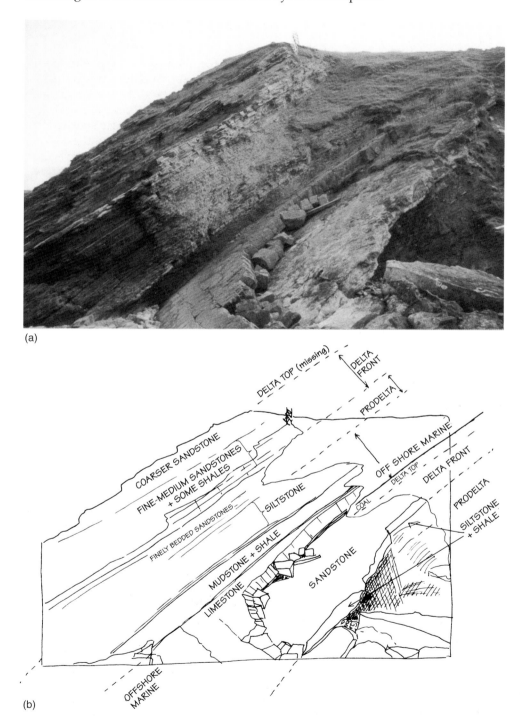

Figure 8.9 (a) Two incomplete Carboniferous cycles in Northumberland, consisting of limestone, mudstone, siltstone and sandstone. The lower of the two cycles is topped by a thin band of coal.
(b) Field sketch of the Carboniferous cycle in (a). The gate on the top of the hill is ~1.2 m high.

This type of coarsening-up sequence and change from marine to non-marine conditions is consistent with a prograding delta environment (e.g. the Mississippi and Ganges deltas). Deltas form along coastlines that are dominated by alluvial processes such that the sediment being discharged by rivers into a lake or sea is so plentiful that it cannot be dispersed and instead accumulates into a large splay or fan (Figure 8.10a).

In the Pennines of England, the relative persistence of marine versus deltaic conditions changed over time. During the cycles of the later part of the Early Carboniferous (Late Dinantian), marine limestones, the overlying mudstones, and the deltaic sandstones are of roughly equal prominence, resulting in what is described as a Yoredale cycle. There are at least 20 Yoredale cycles in the Pennines, each of which is slightly different, with some having underdeveloped or missing units. By the Mid-Carboniferous (Namurian), the limestone units in each of the cycles started to become rarer while at the same time, the sandstone units became thicker and coarser. By the Late Carboniferous (Westphalian) marine conditions were a rare event, whereas the delta-top conditions were sufficiently long lived to result in the formation of coal seams, which were until recently of considerable economic value.

Figure 8.10b summarizes three different types of Carboniferous cycle found within northern England. Although the rock types differ in thickness, each cycle represents a general coarsening-up succession, and hence the progradation (building up and out) of a delta. Each sea-level rise (or land subsidence) floods the coal swamp deposits, so that they are overlain by limestone or shale. In time, the delta starts to prograde once more to produce another cycle.

❑ What global process could control the relative sea-level changes needed to produce the series of cycles?

■ The sea-level changes may be partially attributable to the global glacially-driven eustatic sea-level changes (Figure 3.2) that occurred during the Carboniferous.

This would result in laterally continuous cyclic deposits over the whole basin (e.g. northern England into north-western Europe). Although such deposits do occur, localized cyclic deposits are more commonly found. These localized deposits were most likely controlled by the epeirogenic sea-level changes that were associated with the development of the extensional troughs and blocks or simple subsidence.

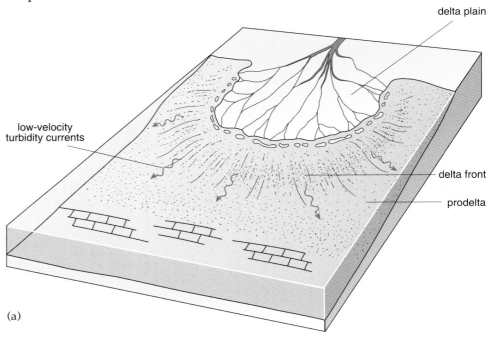

Figure 8.10 (a) Three-dimensional block diagram to show the three main regions of a delta – the delta plain, delta front and prodelta.
(b) Idealized Carboniferous cycles for the Yoredales, Millstone Grit and Coal Measures successions of northern England.
(c) Palaeogeographic sketch of the British Isles showing the south-westward advance of the Carboniferous delta systems in which the cycles formed. (From Kelling, G. and Collinson, J.D. (1992) 'Silesian', in P. Duff and A. J. Smith (eds) *Geology of England and Wales*, Geological Society.)

(a)

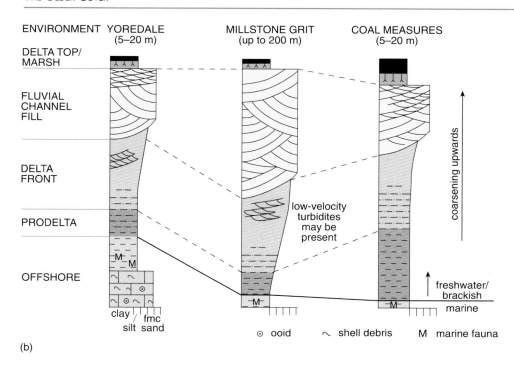

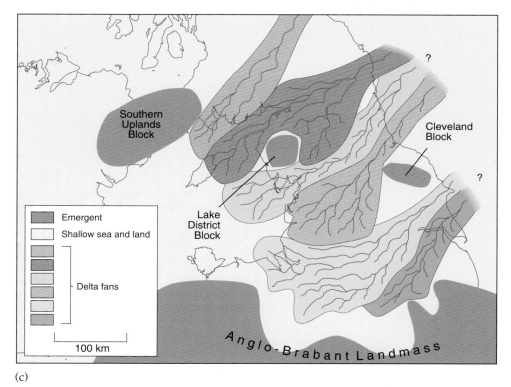

A third potential control involves the rapid switching of the site of deltaic deposition (Figure 8.10c). For example, during a storm, the delta river may burst its banks and change direction, abandoning the original channel network and delta top. Continued subsidence of the abandoned delta top (in response to the weight of sediment on this part of the crust) would cause it to become submerged and result in marine sediments overlying, say, the coal deposit. In reality, a combination of all three models is likely to have controlled the Carboniferous cycles in the British Isles.

8.7 Carboniferous palaeogeography

Plate 6 is a palaeogeographic sketch map of the British Isles at the end of the Early Carboniferous. At this time, the Midland Valley contained a mixed fluvio-lake environment that was subjected to periods of explosive volcanism associated with lithospheric extension and passive continental rifting (Section 8.5). At the same time, northern England was dominated by a series of deltas, prograding south from the ancient uplands. The rest of England, Wales and Ireland was occupied by a series of basins characterized by shallow-water marine conditions. In these basins, shallow-water dark grey mudstones and **bioturbated**, fine-grained fossiliferous, **micritic** limestones were deposited, typical of a low-energy environment. These limestones formed at the boundary between a shallow-water shelf environment and deeper basinal waters, similar to carbonate reefs and platforms found in the present-day Bahamas.

Similar conditions continued into the Mid-Carboniferous (Namurian, Plate 7) as the global sea-level continued to rise. By the end of the Mid-Carboniferous, a marine regression (sea-level fall) had allowed fluvial and deltaic systems to advance more fully into the basins (Figure 8.10c). As the deltas advanced, thick units of coarse-grained and mineralogically mature cross-stratified sandstones were deposited in the fluvial channels that fed the delta systems (e.g. Millstone Grit, Figure 8.11).

By the Late Carboniferous (Westphalian), the Coal Measures Series was deposited across the Midland Valley, northern and central England, south-east Ireland, southern Wales, and into central Europe (Figure 8.12 and Plates 8–9). With time, later Coal Measures deposits began to incorporate progressively more red sediments (Barren Red Series), indicative of slightly more arid conditions in contrast to the waterlogged swamps of the earlier coal deposits (Plate 9). The large temporal and spatial distribution of the Coal Measures 'swampy' environment can be attributed to a fine balance between subsidence rate and sediment input from the Caledonian highlands, the Anglo–Brabant landmass and the Variscan highlands to the south.

To the south of the Anglo–Brabant landmass, all the coalfields were affected by the Variscan Orogeny, subjecting the bituminous coals to elevated temperatures and converting them to higher-grade **anthracite** (i.e. coal with a higher carbon and lower volatile content). To the north of the landmass, some of the Coal Measures deposited in the central and southern North Sea basins eventually became deeply buried beneath thick Younger Cover. This burial and maturation of the organic material led to the formation of some of the major natural gasfields in the British Isles (Figure 8.12).

Figure 8.11 Well-developed cross-stratification in one of the Millstone Grit cycles found across the Peak District in northern England.

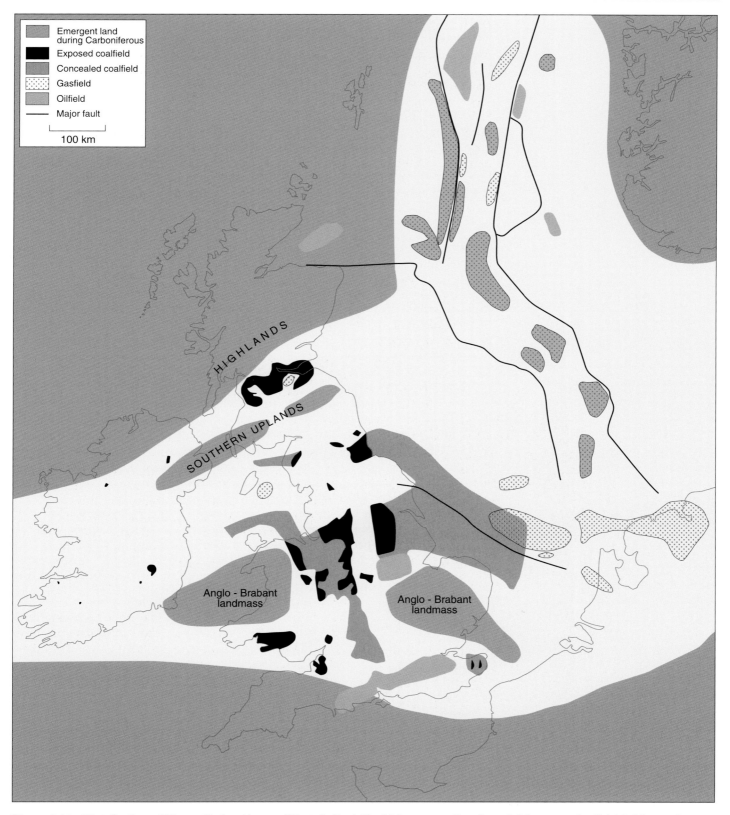

Figure 8.12 Distribution of Upper Carboniferous (Westphalian) Coal Measures, oil and gasfields across the British Isles and mainland Europe. Note that in the northern North Sea, some of the oil and gasfields are Jurassic in age, as is the oilfield of southern England.

Therefore, during the Carboniferous, four dominant themes affected the palaeogeography of the British Isles, namely:

(1) A marine transgression during the Early Carboniferous.

(2) Lithospheric extension and isostatic readjustment of the crust, forming subsiding troughs on either side of upland blocks. This extension led to passive continental rifting and associated volcanic activity in the Midland Valley.

(3) Progressive marine sedimentation over the slowly subsiding blocks and troughs.

(4) Cyclical sedimentation produced by advancing and retreating deltas, associated with eustatic (probably glacial-driven) and epeirogenic sea-level changes.

8.8 Summary

Devonian Period

- The Devonian was characterized by arid conditions, with the British Isles straddling 15°–20° S.
- North of Devon, the major part of the British Isles was emergent throughout the whole of the Devonian Period and was subject either to rapid rates of erosion or to periods of non-deposition. Where continental Old Red Sandstone deposits have accumulated, they can be seen to be indicative of alluvial, fluvial and/or lake environments. To the south of Devon, the British Isles was characterized by marine conditions, with shallow to deep-water ocean-basin sediments forming at the edge of the Rheic Ocean.
- Two distinct stages of igneous activity can be recognized in the Devonian. The first was associated with the end of the Caledonian Orogeny and collision of Avalonia and Laurentia, with the emplacement of post-orogenic granitic plutons as a result of partial melting of the thickened continental crust. The second type of intrusive and extrusive igneous activity cannot be spatially or temporally related to subduction or collision associated with the closure of the Iapetus Ocean. Instead, this activity relates to localized lithospheric extension, associated with thermal relaxation of the crust after the Caledonian Orogeny and strike–slip movement along major faults (such as the Highland Boundary and Great Glen Faults).

Carboniferous Period

- The Carboniferous was marked by a series of eustatic sea-level rises, with the British Isles also experiencing significant climatic changes, going from arid to semi-arid and then monsoonal conditions associated with a northwards drift across the equator, as well as increasing global temperatures.
- A combination of eustatic and epeirogenic sea-level rises resulted in a marine transgression across the British Isles, initially forming shallow-water shelf limestones, which were superseded by cyclic deposition of limestone, shale, sandstone and coal.
- At the same time, the lithospheric extension that had commenced in the Devonian continued into the Early Carboniferous, resulting in the formation of a series of upstanding blocks and intervening troughs across the British Isles. Associated with this was localized shallow-level intrusive and extrusive igneous activity, which was especially prevalent in the Midland Valley.

- Over time, the upland blocks (and larger landmasses) were subjected to erosion, with this material deposited in the adjacent troughs. This altered the neutral buoyancy of both blocks and troughs, causing the blocks to remain buoyant while the troughs subsided. This had a profound effect on the timing and type of sediments that accumulated in the troughs, as well as on top of the blocks.
- By the Mid-Carboniferous, rifting associated with extension ceased, resulting in the blocks and troughs subsiding together as the crust continued to undergo thermal relaxation, forming large sag-basins.
- By the end of the Late Carboniferous, climatic conditions were beginning to change once again, with red beds, indicative of more arid conditions, interbedded with other Upper Coal Measures lithologies.

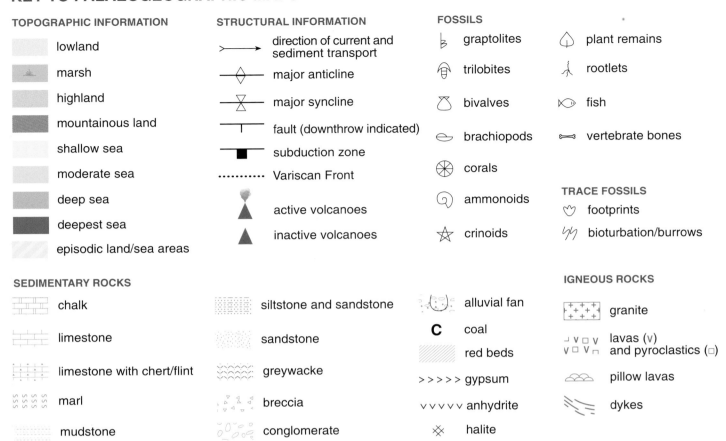

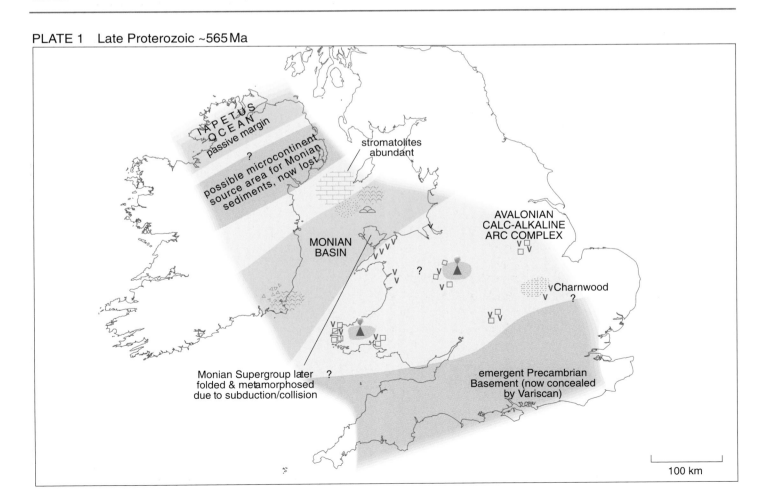

PLATE 1 Late Proterozoic ~565 Ma

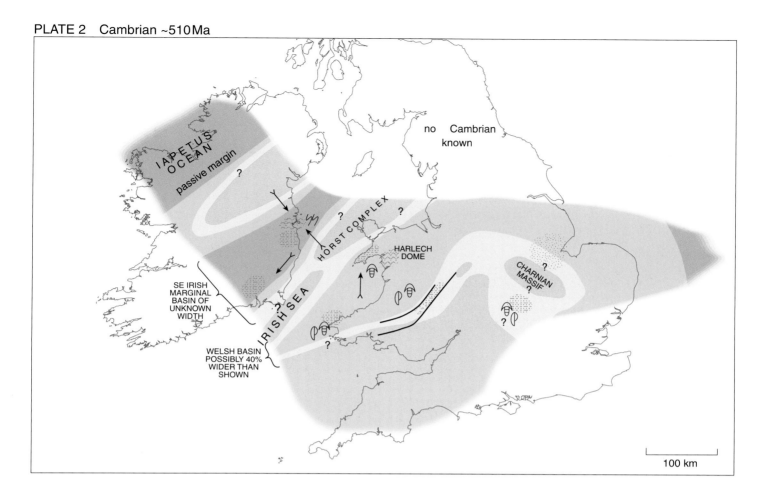

PLATE 2 Cambrian ~510 Ma

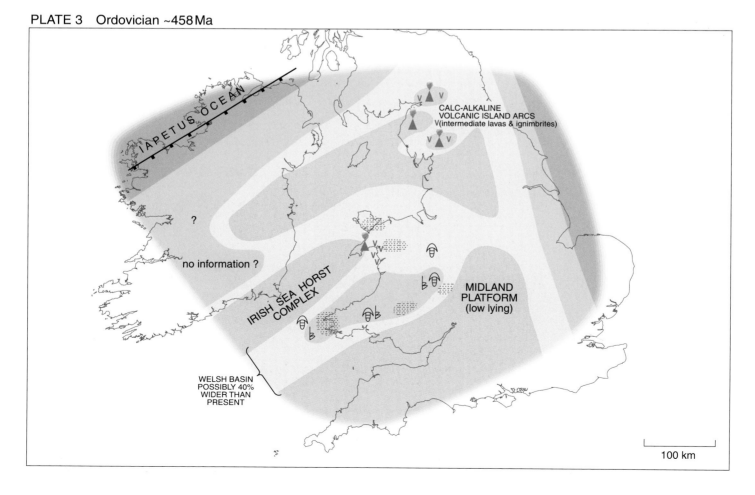

PLATE 3 Ordovician ~458 Ma

PLATE 4 Early Devonian ~400 Ma

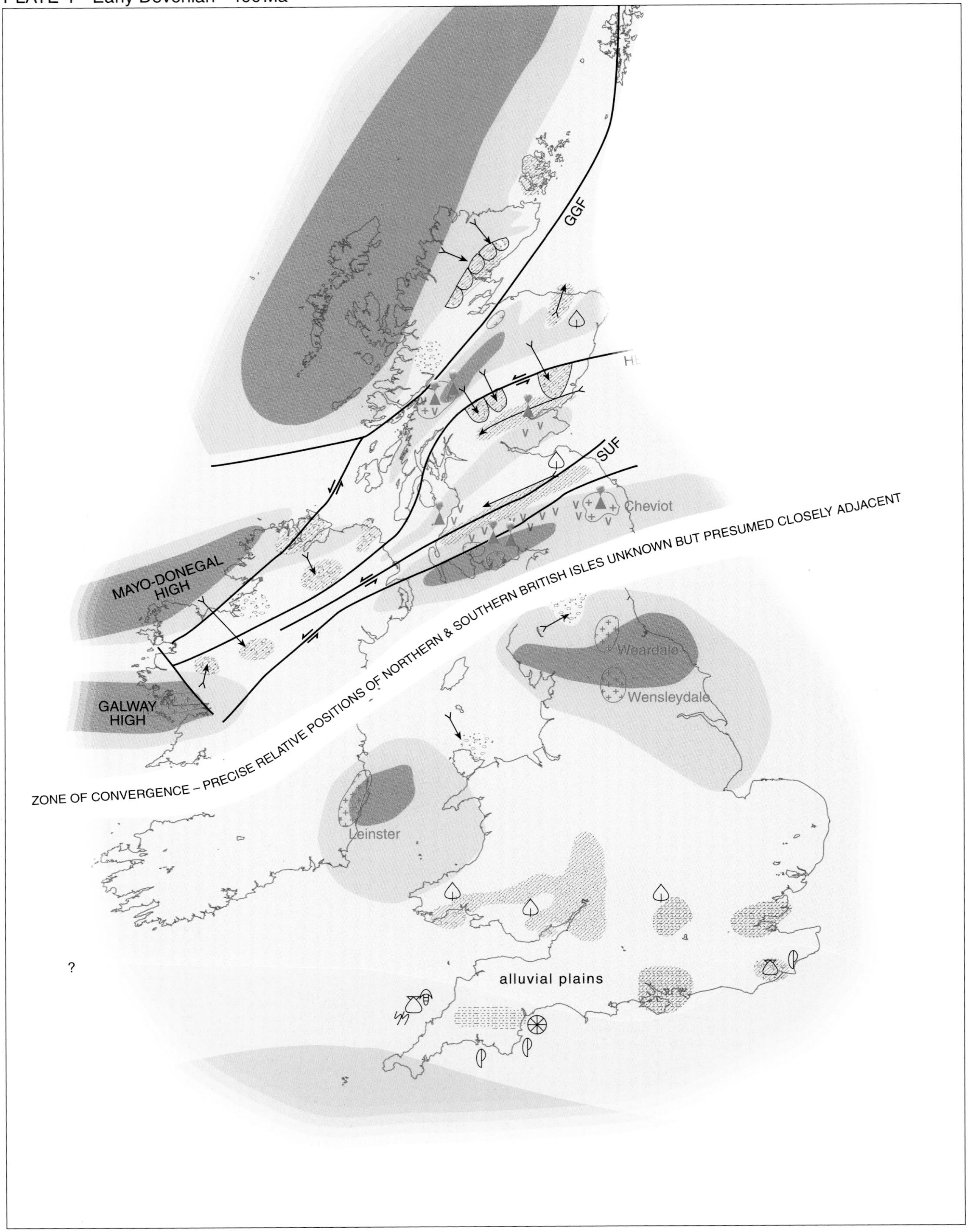

PLATE 5 Late Devonian ~365 Ma

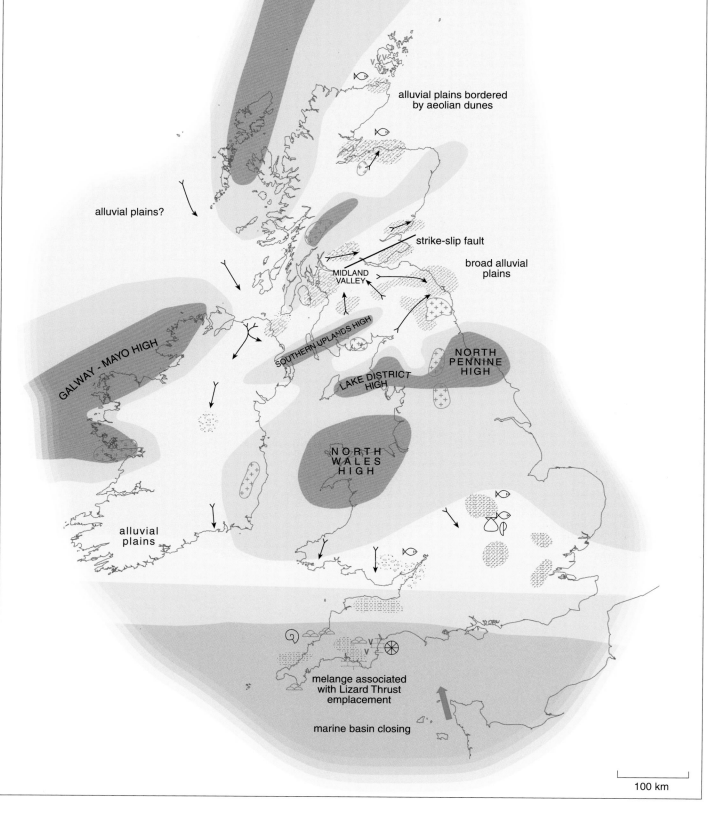

PLATE 6 Early Carboniferous (Dinantian) ~337 Ma

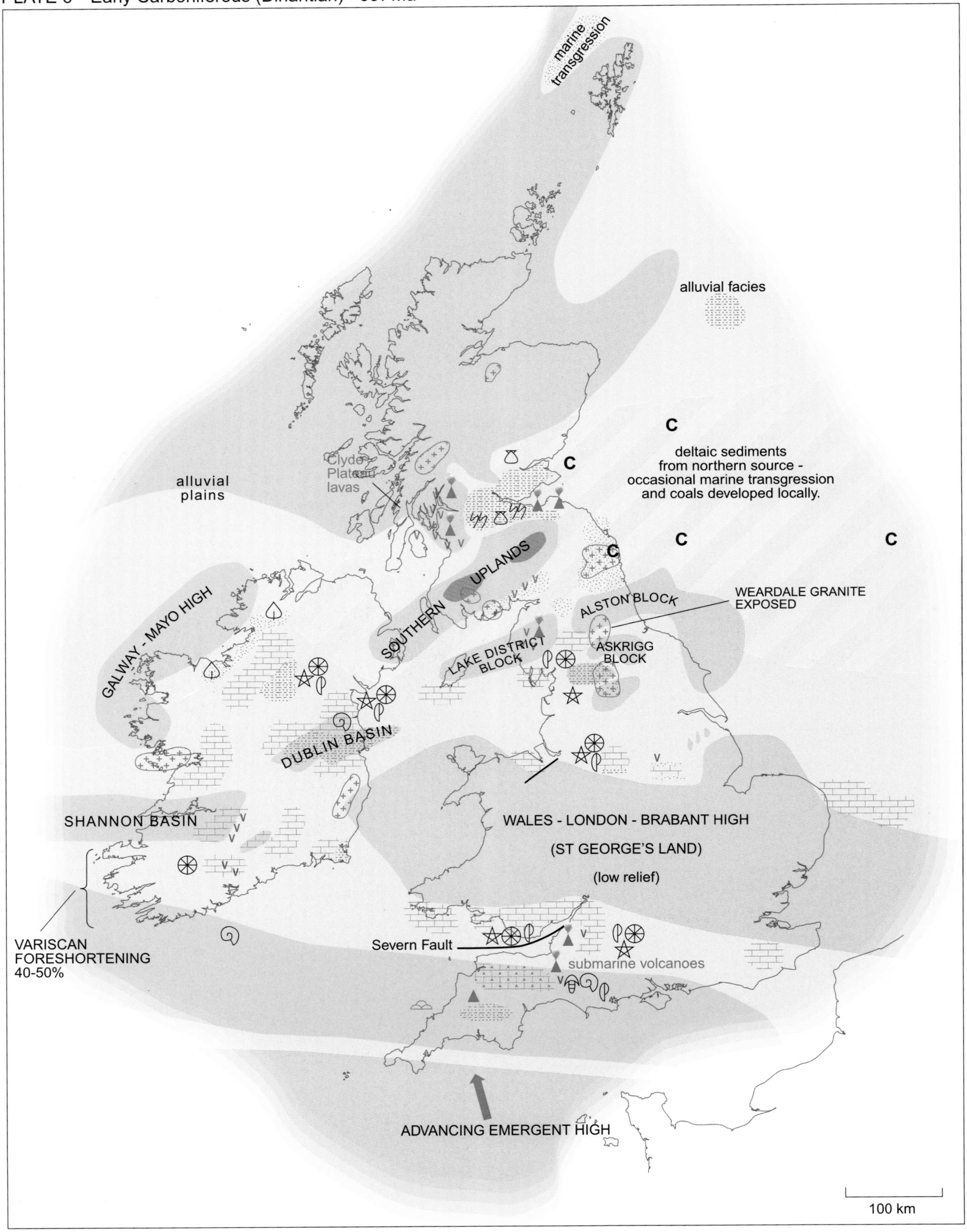

marine transgression

alluvial facies

deltaic sediments from northern source - occasional marine transgression and coals developed locally.

C C C C
C

alluvial plains

Clyde Plateau lavas

SOUTHERN UPLANDS

GALWAY - MAYO HIGH

WEARDALE GRANITE EXPOSED

ALSTON BLOCK

LAKE DISTRICT BLOCK

ASKRIGG BLOCK

DUBLIN BASIN

SHANNON BASIN

WALES - LONDON - BRABANT HIGH

(ST GEORGE'S LAND)

(low relief)

Severn Fault

submarine volcanoes

VARISCAN FORESHORTENING 40-50%

ADVANCING EMERGENT HIGH

100 km

PLATE 7   Mid-Carboniferous (Namurian) ~325 Ma

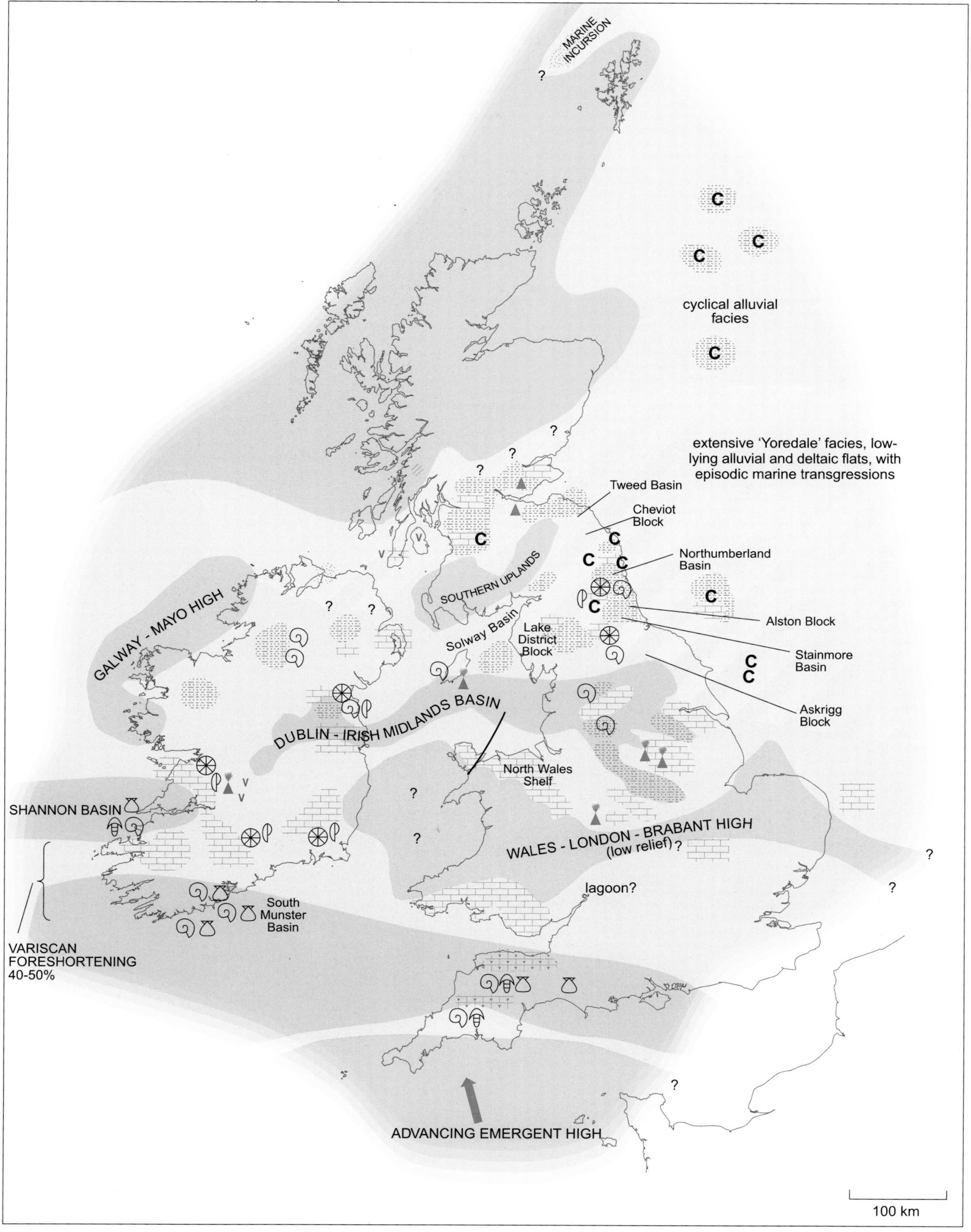

MARINE INCURSION ?

C C C
C
cyclical alluvial facies
C

extensive 'Yoredale' facies, low-lying alluvial and deltaic flats, with episodic marine transgressions

Tweed Basin
Cheviot Block
C
C C   Northumberland Basin
C
C   Alston Block

SOUTHERN UPLANDS

Solway Basin
Lake District Block
Stainmore Basin

GALWAY - MAYO HIGH

C C
Askrigg Block

DUBLIN - IRISH MIDLANDS BASIN

North Wales Shelf

SHANNON BASIN

WALES - LONDON - BRABANT HIGH (low relief) ?

lagoon?   ?

?

South Munster Basin

VARISCAN FORESHORTENING 40-50%

ADVANCING EMERGENT HIGH

100 km

PLATE 8 Late Carboniferous (Westphalian) ~309 Ma

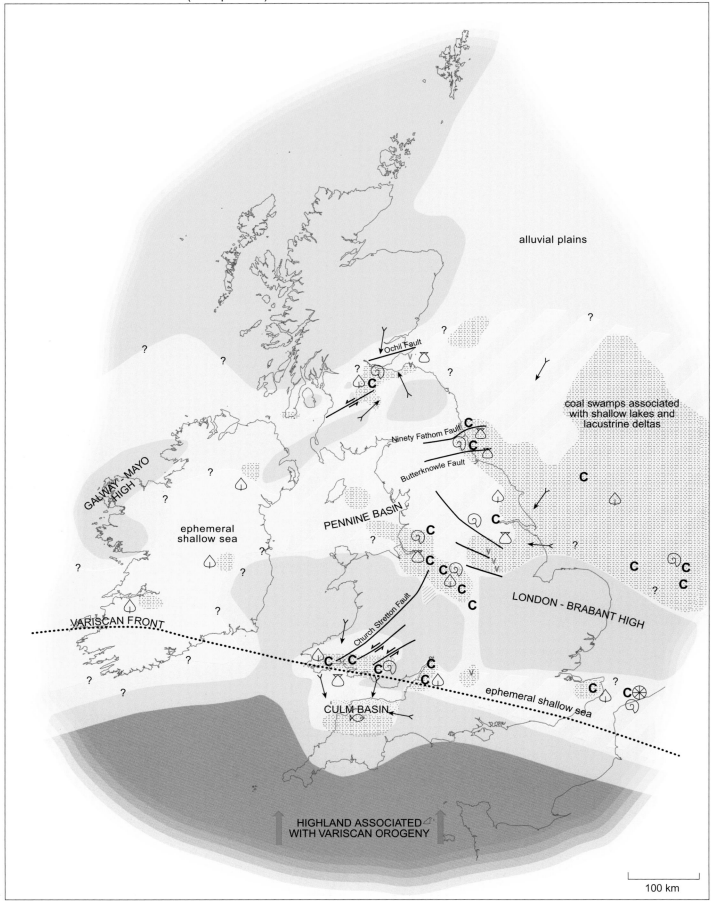

PLATE 9   Late Carboniferous (Late Westphalian) ~304 Ma

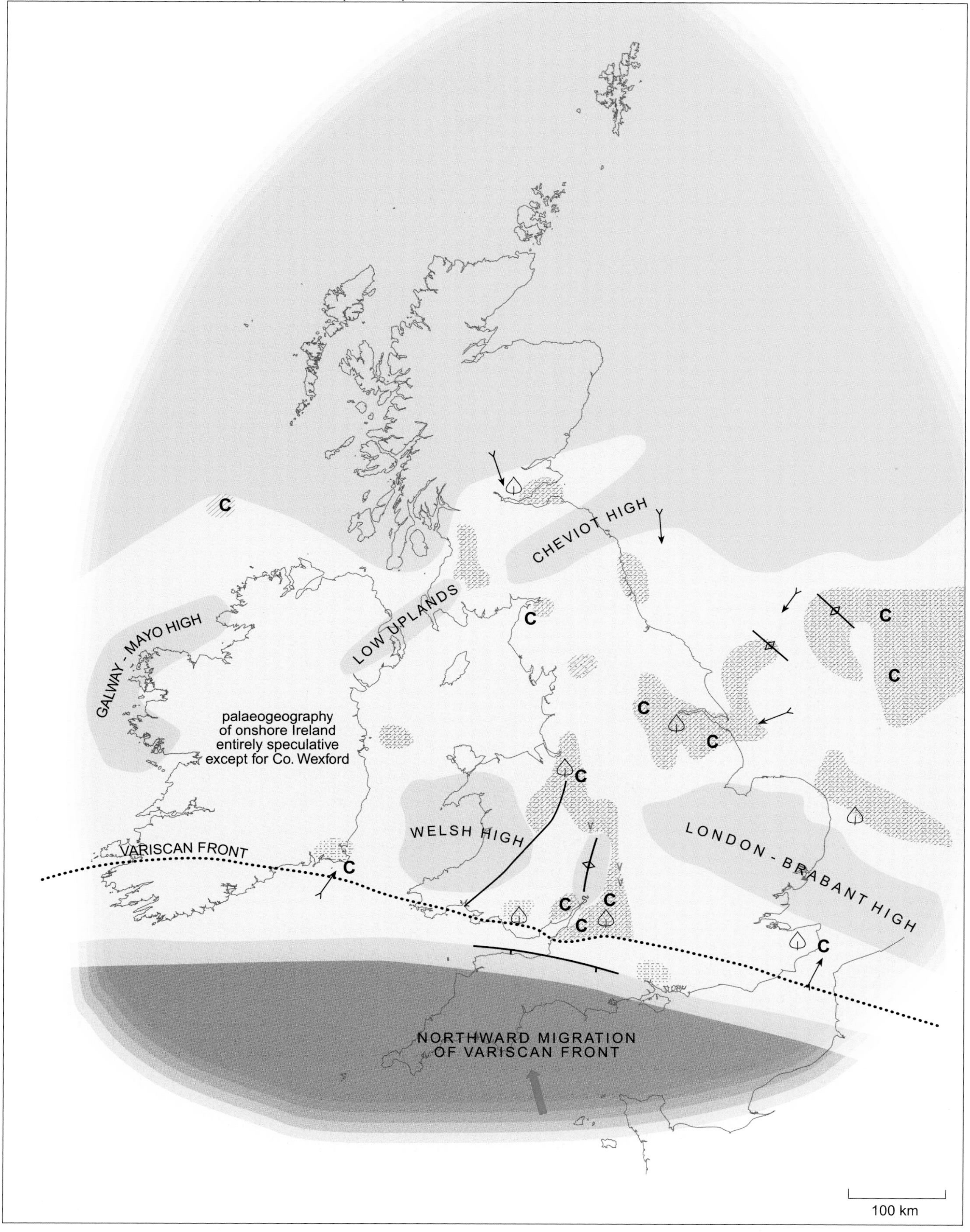

PLATE 10   Late Permian ~255 Ma

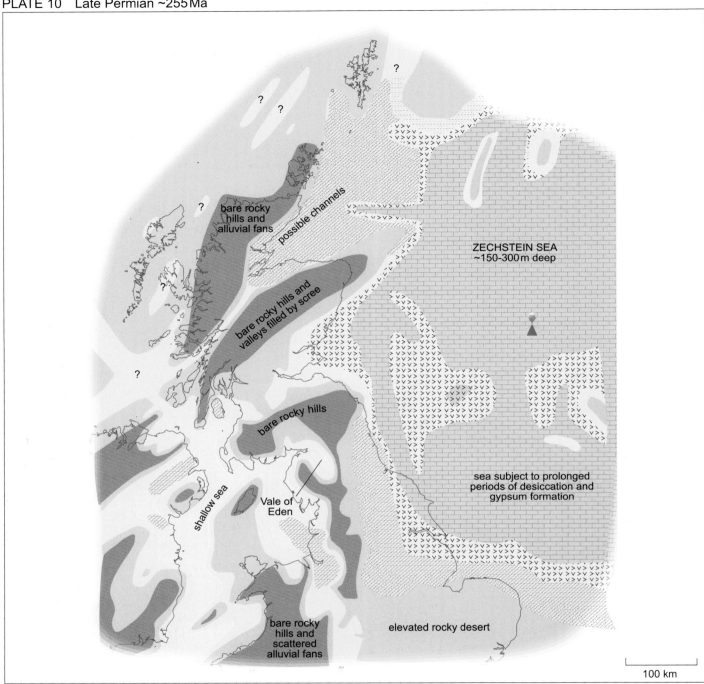

PLATE 11 Mid Triassic ~235 Ma

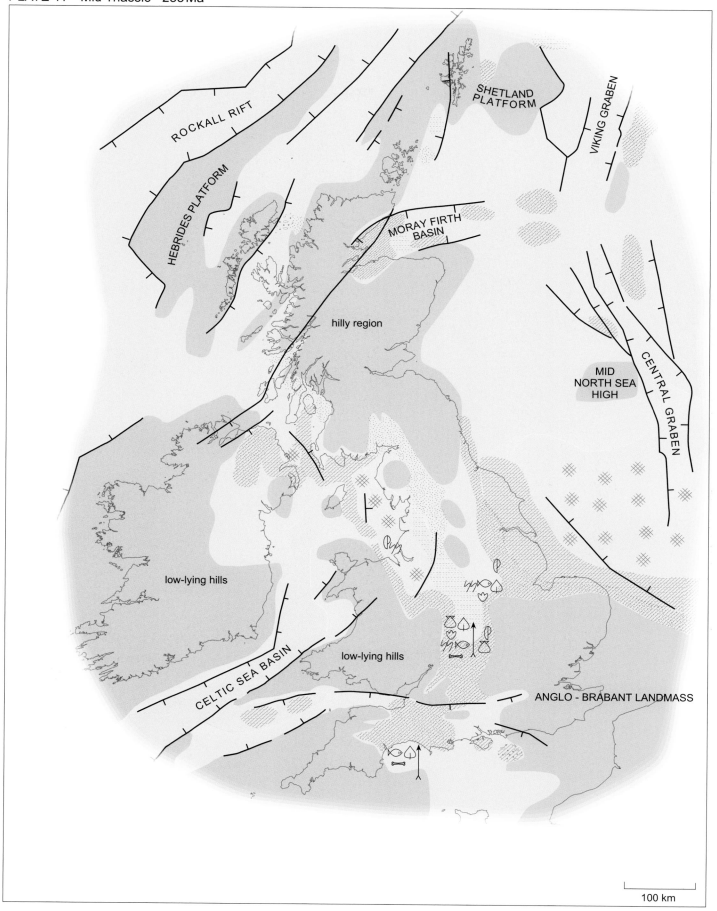

PLATE 12 Early Jurassic ~204 Ma

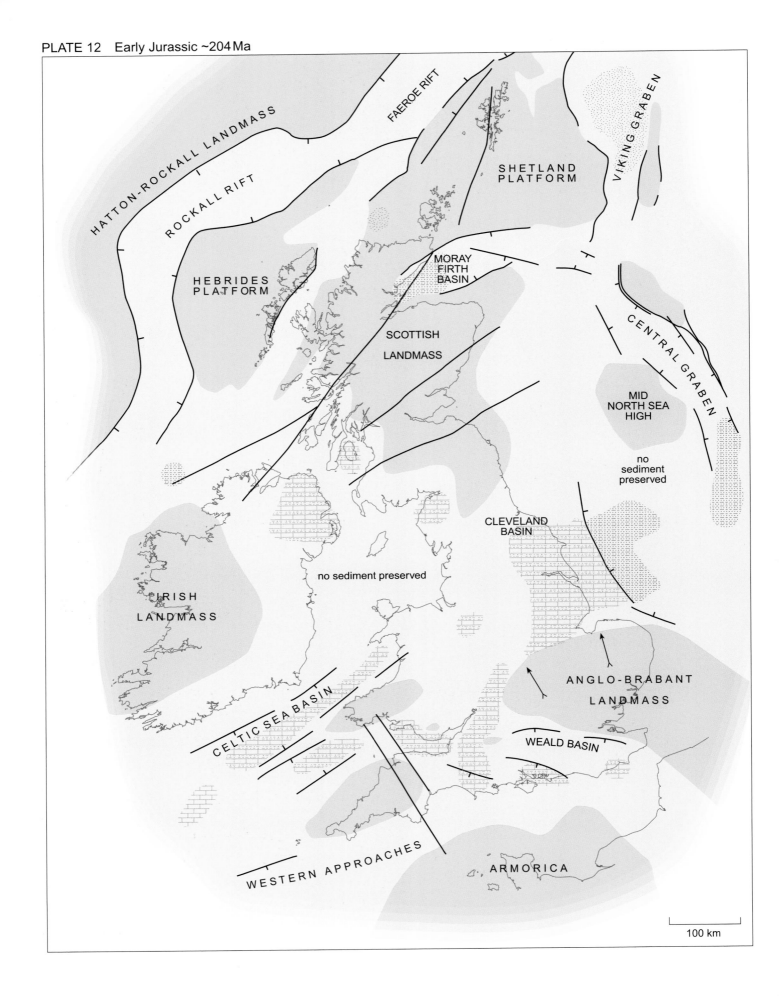

PLATE 13   Mid Jurassic ~175 Ma

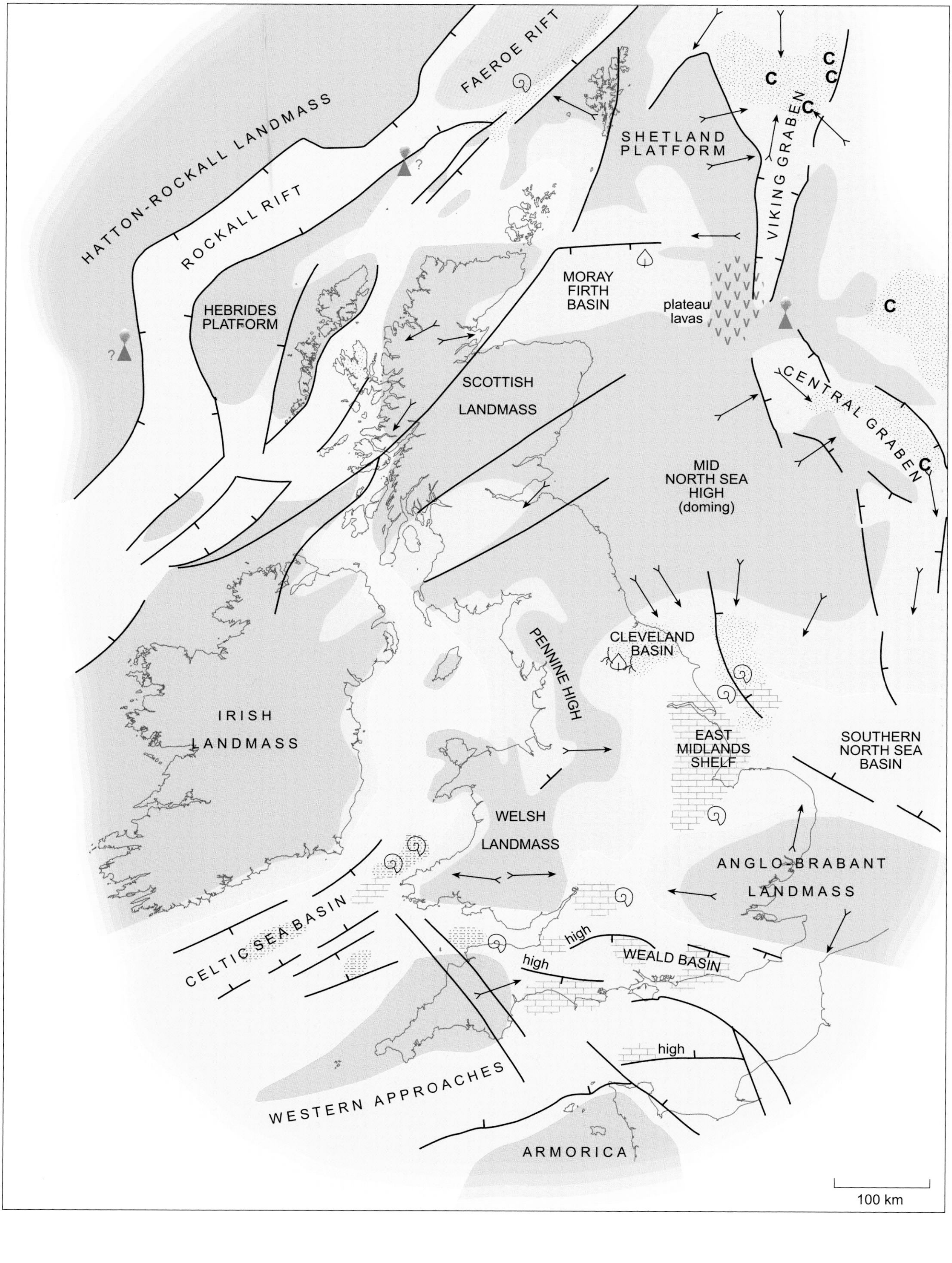

PLATE 14   Late Jurassic ~149 Ma

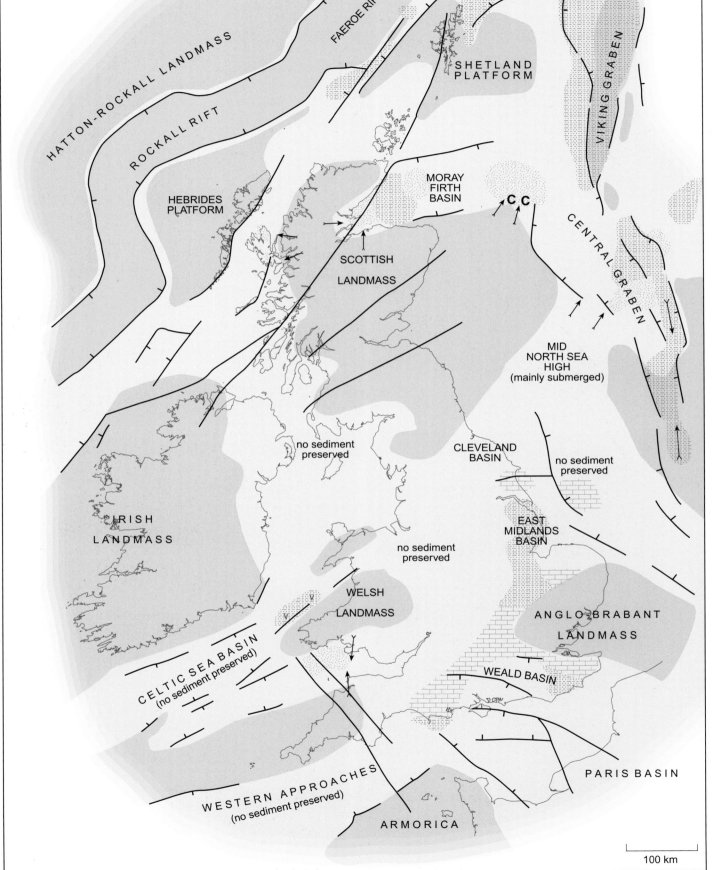

PLATE 15   Early Cretaceous ~97 Ma

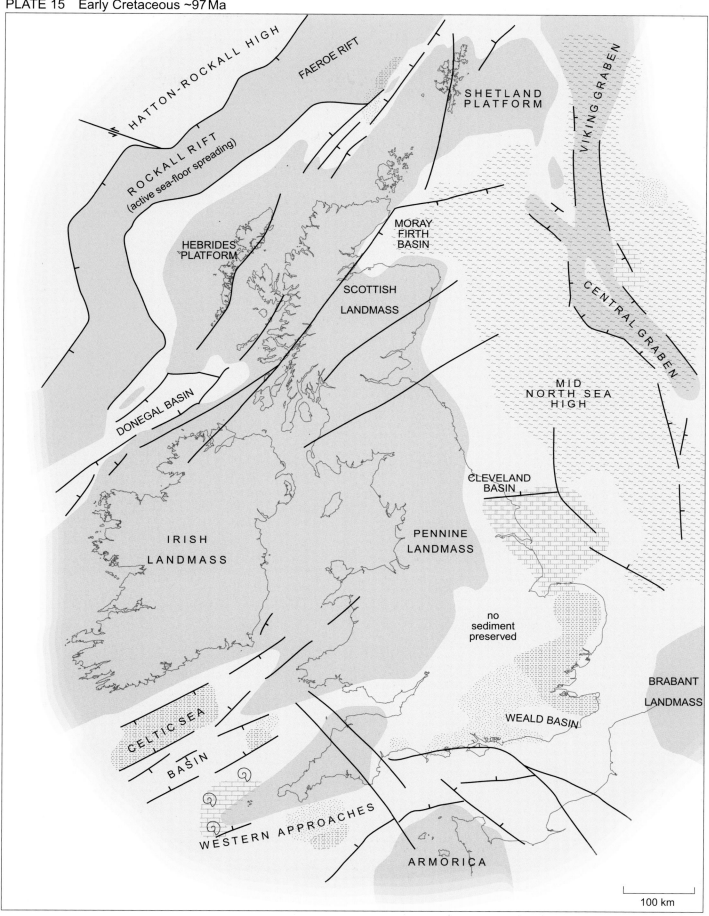

PLATE 16   Late Cretaceous ~75Ma

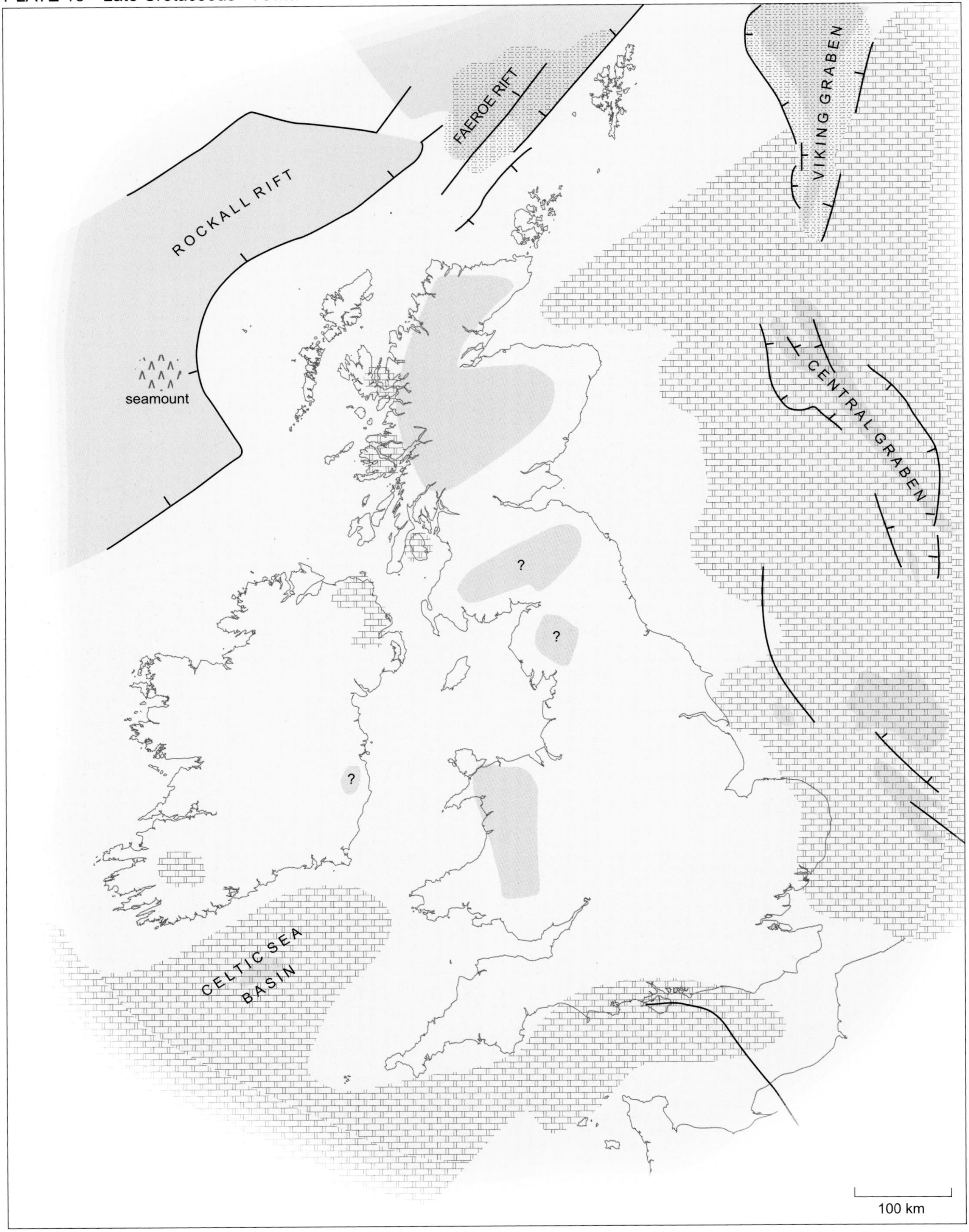

ROCKALL RIFT

FAEROE RIFT

VIKING GRABEN

CENTRAL GRABEN

seamount

CELTIC SEA BASIN

100 km

PLATE 17 Tertiary ~60 Ma

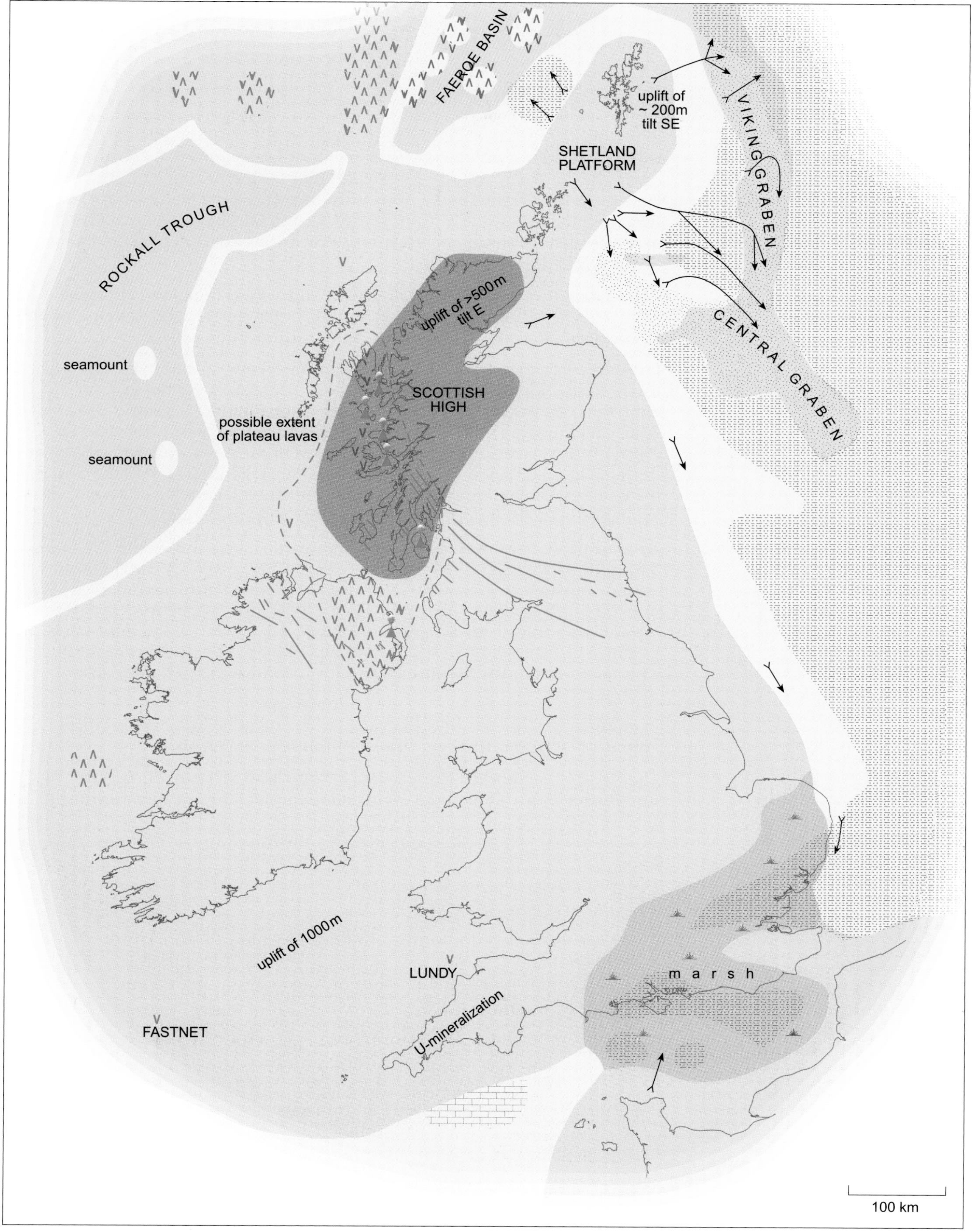

ROCKALL TROUGH

FAEROE BASIN

SHETLAND PLATFORM

uplift of ~200 m tilt SE

VIKING GRABEN

CENTRAL GRABEN

seamount

seamount

SCOTTISH HIGH

uplift of >500 m tilt E

possible extent of plateau lavas

uplift of 1000 m

LUNDY

U-mineralization

FASTNET

marsh

100 km

9 The Variscan Orogenic Belt

9.1 Introduction

In Section 8, you read about the range of generally undeformed rocks of the Older Cover along with their environments of formation and used this to establish how the British Isles evolved during the Devonian and Carboniferous Periods. In this Section, you will be asked to examine the deformed rocks and structures that formed as a result of the Variscan Orogeny, and establish how this event affected the structural and lithological geology of the Older (and Younger) Cover across the British Isles.

> Question 9.1 Which part(s) of the British Isles has been most strongly affected by deformation and metamorphic events associated with the Variscan Orogeny? (Refer back to Section 5.5 if you cannot remember.)

Looking beyond the British Isles, the Variscan Orogenic Belt can be followed south-westwards across the Atlantic to New Brunswick and the eastern American Atlantic coast, and south into northern France, Belgium, Germany, Spain and Portugal, where the effects of the orogeny are most strongly felt (Figure 9.1). This region consists of a series of terranes that collided sequentially during latest Silurian to Late Carboniferous times, associated with the closure of the Rheic Ocean between Gondwana and Laurentia (Figure 3.1). These events assembled western Europe into roughly its present configuration.

An ancient ocean existed to the south of the British Isles during the Palaeozoic, and evidence for this can be found near Lizard Point (SW(10)71). This is an ophiolite complex (radiometrically dated to ~400 Ma, Early Devonian), representing a section of obducted oceanic crust (basalts–gabbros) and upper mantle (serpentinized peridotite, see Sections 4.2.4 and 5.5). Secondly, the large granite masses in south-west England could either be the product of subduction associated with ocean closure or formed by partial melting of collisionally thickened crust.

In addition to these two lines of evidence, the intense folding in the Upper Palaeozoic strata in this region is also indicative of collision.

The Rheic Ocean formed as a result of lithospheric extension during the Cambrian to Early Ordovician. This extension rifted the northern margin of

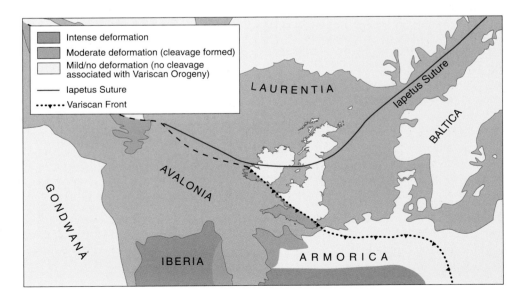

Figure 9.1 Schematic palaeogeography of the Late Carboniferous (Westphalian) terranes on either side of the present day North Atlantic Ocean, showing the extent of the Variscan Orogenic Belt. The occurrence of Variscan rocks on either side of the Atlantic Ocean supports the concept that the continents bordering the ocean were once united. (From Glover, B. W. *et al.* (1996) 'A second major fluvial sourceland for the Silesian Pennine Basin of northern England', *J. Geol. Soc.* Vol. 153.)

Gondwana apart, to form the micro-continent of Avalonia which contained the southern British Isles and northern edge of mainland Europe (Figure 3.1a–b). Other micro-continental terranes rifted away from Gondwana at this time include Armorica (containing present day Brittany to central Germany) and Iberia (present day Spain and Portugal), as well as several others.

At the same time as the Rheic Ocean was opening, the Iapetus Ocean situated to the north was beginning to close because of subduction below the northern margin of Avalonia and the southern margin of Laurentia (Figures 3.1b–c and 7.12). With the closure of Iapetus and the Caledonian Orogeny, Avalonia united with Laurentia (and Baltica) to form one large continent (Figure 3.1d), which is often referred to as the Old Red Sandstone (ORS) landmass.

Avalonia was the first of the former Gondwana terranes to collide with Laurentia, but all the rest followed when the Rheic Ocean stopped spreading and started to close. Armorica may have impinged upon Avalonia by the end of the Silurian, but usual evidence of subduction is largely absent and so the collision was probably oblique. However, subduction proceeded further to the south and Iberia ran into Armorica, causing a back-arc basin to open up between Avalonia and Armorica within which the oceanic lithosphere now represented by the Lizard ophiolite (Plate 5) was formed. By the beginning of the Permian, the Rheic Ocean had completely closed, uniting the ORS landmass with Gondwana and all the intervening terranes to form the vast super-continent of Pangea (Figure 3.1f).

Throughout the rest of this Section, we will examine the nature of this ocean closure, the style of continental collision and how the resultant Variscan Orogeny affected the geology of the British Isles. Deformation and metamorphism are more intense within Armorica and the terranes to its south, but these do not form part of our story.

9.2 Recognizing Variscan deformation in the rock record

9.2.1 Lithological diversity in south-west England

In Section 8, the variety of Devonian and Lower Carboniferous sediments found in south-west England between north Devon and Cornwall, was described. In north Devon, the Upper Palaeozoic consists of alternating alluvial to fluvial and shallow-water marine siliciclastics that grade southwards into shallow-water marine fossiliferous limestones, including limestone turbidites, representing the slope deposits found at the front of carbonate platforms (Figure 9.2 overleaf). Continuing south, in north Cornwall this shallow-water marine sequence prograges into a series of turbiditic sediments containing volcanic breccias and pillow lavas (now all metamorphosed to slate), joined by the deposition of chert in the Early Carboniferous. Chert forms in deep basins, and is characteristic of very low sedimentation rates and hence a low-energy environment. This continental to shallow and then deep-water marine succession in Devon and Cornwall is indicative of a subsiding continental shelf and slope environment that passes into a progressively deeper ocean basin (Figure 9.2).

By the Late Carboniferous, eustatic sea-levels were beginning to fall (Figure 3.2). This marine regression resulted in the shallow-water continental shelf and slope deposits moving southwards across south Devon and north Cornwall.

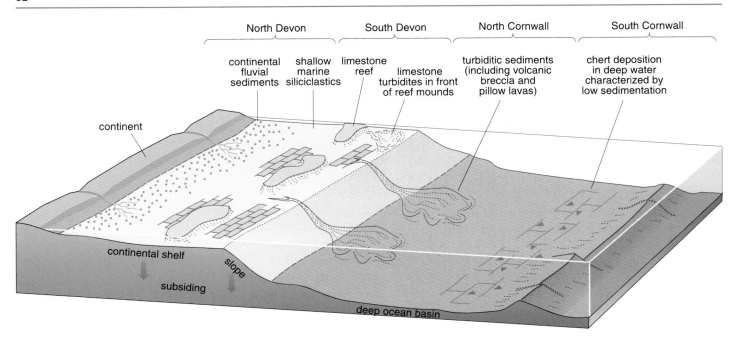

Figure 9.2 Depositional environments across south-west England during the Devonian and Early Carboniferous, prograding from the edge of a continental landmass in north Devon to a deep ocean basin in south Cornwall.

A similar Devonian to Lower Carboniferous succession can be recognized across south-west Ireland, where a succession of alluvial and fluvial continental sediments grades into a carbonate platform and deep-water marine siliciclastics and turbidites in Counties Kerry and Cork.

One final rock type that is abundant in south-west England is granite, forming a series of intrusions that trend south-westwards from Dartmoor to the Isles of Scilly. The Land's End to Bodmin Moor granites have a maximum age of Late Devonian, whilst the Dartmoor granite is of Late Carboniferous age. As no contact relationship is visible between the granites and Younger Cover, a minimum age of emplacement cannot be obtained.

A more accurate age of emplacement of ~280–270 Ma (Early Permian) has been obtained for all of the granites by U–Pb radiometric dating. Furthermore, geophysical evidence reveals that below the surface, the separate granites are actually one continuous sheet-like plutonic mass, which has a very irregular and knobbly surface topography. (No equivalent-aged granites are found in the Variscan Orogenic Belt in Wales or Ireland.)

Petrographic, structural and field-based studies carried out on these granites show that they are not deformed. This fact combined with their Early Permian emplacement age implies that the granites were intruded after the surrounding Upper Palaeozoic sediments had been deformed by the Variscan Orogeny. In other words, the granites are post-orogenic. Using this age-related evidence, the final stage of Variscan deformation must therefore have occurred *after* the Upper Carboniferous (Westphalian) sediments were deposited, but *before* the Early Permian granites were intruded.

Therefore, to summarize so far, the Rheic Ocean separated the British Isles and northern mainland Europe from the rest of Europe during most of the Palaeozoic. Towards the end of this Era, the Rheic Ocean closed, with the collision of Gondwana terranes, and ultimately Gondwana itself, with Laurentia. Continental collision resulted in deformation and metamorphism, followed by the emplacement of post-orogenic granites.

9.2.2 Variscan deformation styles

Within the British Isles, the northern limit of strong Variscan deformation is marked by the Variscan Front (Figure 5.2). This is not a sharp line but represents the approximate boundary between regions to the south that underwent metamorphism and/or strong deformation during this event from areas to the north that were not metamorphosed and only weakly deformed (Plate 8).

Variscan deformation was caused by N–S crustal shortening. This produced a series of asymmetrical folds with E–W trending fold axes across the southern British Isles, e.g. the South Wales Coalfield, which is a broad E–W syncline, with closure at the eastern end, indicating a basin-like structure. Steeper dips on the southern limb of this syncline reflect the northwards-directed Variscan movements. Another E–W syncline occurs in south-west England. The steeper southern limbs of synclines indicate that the dominant direction of shortening was from the south. Furthermore, on the basis of outcrop width and relative distance between the fold axes, the intensity and complexity of Variscan deformation can be seen to *increase* to the south (Figure 9.3).

However, south of St. Austell, the whole of the Devonian is characterized on Figure 9.3 by a series of E–W trending thrust planes rather than fold axes. This reflects the structural complexity of the area south of St. Austell, which consists of a series of intensely folded and thrusted sections of crust (and upper mantle), composed of turbiditic greywackes, conglomerates (containing igneous, metamorphic and sedimentary clasts) and minor volcanics. Further to the south and south-east, two distinct thrusted sections can be identified, consisting of the Lizard Point ophiolite complex (SW(10)6516–8223) and the metamorphosed schists near Start Point (SX(20)83), both of which are Devonian in age. (The metamorphic gneiss and schist at Start Point and the Lizard Point ophiolite were once thought to be Precambrian; these have now been reassigned to the Mid-Devonian (~400 Ma).)

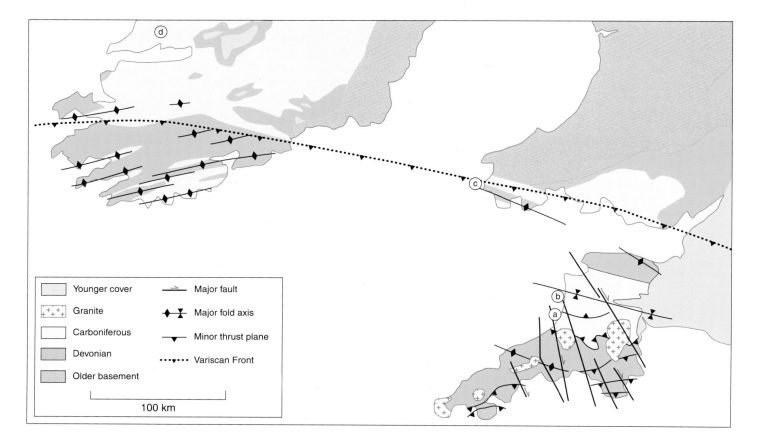

Figure 9.3 Simplified geological map of the south of Ireland, south Wales, Devon and Cornwall, showing the scale of Variscan folding, faulting and thrusting. (a)–(d) show the locations of the photographs in Figure 9.6.

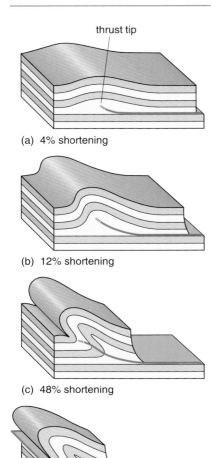

Figure 9.4 Block diagrams showing stages in the development of a fold pair related to a thrust tip.
(a) Ductile thickening ahead of the thrust tip causes folds to form in the hangingwall.
(b) This leads to asymmetric folds forming, with the sense of overturning consistent with the direction of hangingwall movement.
(c) The thrust propagates through the overturned limb of the fold.
(d) The anticline in the hangingwall is carried forward as the thrust tip advances beyond the fold.

In the simplest terms, these thrusted sections along with many other smaller examples found across the south-west of the British Isles have formed as a result of crustal shortening, to form thrust **tip-related folds**. Figure 9.4 illustrates the steps involved in this mechanism of folding and thrusting. As a succession of sub-horizontal undeformed sediments (like those originally in Devon and Cornwall) undergoes shortening parallel to the strata layers, differences in their competence eventually lead to the formation of a low-angle thrust (Figure 9.4a). Ductile thickening occurs at the tip of this thrust, as competent and incompetent layers deform in response to shortening. As the intensity of shortening increases, the extent of crustal thickening also increases, leading to the formation of an asymmetric fold above the thrust tip (Figure 9.4b–c). Eventually, the extent of crustal shortening is sufficiently high to allow the thrust to propagate through the overturned fold, carrying the **hangingwall** forward as the thrust tip advances beyond the original fold axis (Figure 9.4d).

In general, folds that form in front of thrust tip-planes are asymmetrical, with inclined axial planes (dipping towards the direction from which thrust movement came) and one limb overturned or at least much more steeply dipping than the other. This is true of the larger-scale E–W trending folds in Cornwall and south Devon, and on a smaller scale there are many fault-bend folds too, where the hangingwall has been transported over a **footwall** ramp

9.2.3 Deformation and emplacement history of the Lizard ophiolite complex

Figure 9.5 summarizes the deformation and emplacement history of the Lizard ophiolite complex. This consists of a series of folded and thrusted 'slices' through the oceanic crust and mantle, formed by the thrust tip method described above.

At least four stages of deformation and thrusting have been identified in the Lizard ophiolite complex, stacking slices of deeper oceanic crust (and subsequently upper mantle) on top of shallower oceanic crustal layers (Figure 9.5a–b). With each deformation stage, earlier thrust slices were refolded and thrust further northwards (Figure 9.5c). This led to a progressive over-thrusting and thickening of the crust, increasing the metamorphic grade (initially forming schists then gneiss), and eventually resulting in small-scale partial melting, with the melts migrating along the thrust planes. When shortening due to collision ceased and the metamorphic conditions waned, the thickened thrusted ophiolite complex collapsed down into its current configuration, as a result of thermal relaxation and isostatic re-equilibration with the surrounding crust (Figure 9.5d).

9.2.4 Variscan folding

Moving north from St. Austell, the intensity of deformation, metamorphism and shortening progressively decrease. For example, at Millhook Haven (SS(21)1800) in north Cornwall, structural analyses on the folds indicate that ~60% crustal shortening has occurred (Figure 9.6a). The strongly metamorphosed Carboniferous strata here are deformed into **recumbent chevron folds**, with a strong slaty **cleavage**. A few kilometres north of this site, the amount of shortening decreases to ~50%, forming less intense, but more irregular folds, which exhibit a moderate to good slaty cleavage (Figure 9.6b). This pattern of decreasing deformation and metamorphism continues northwards into south Wales, where folds equivalent to ~10% crustal

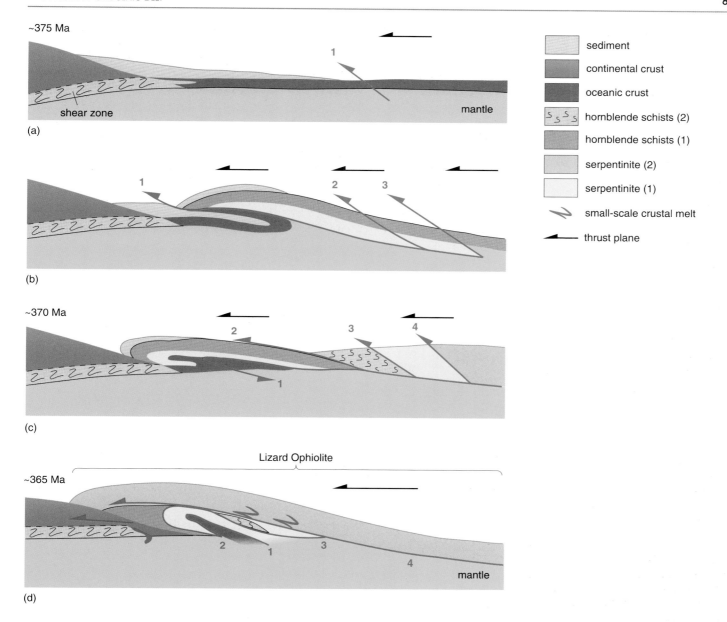

Figure 9.5 Deformation and emplacement history of the Lizard ophiolite.
(a) Thrusting is initiated ~ 375 Ma, after an initial period of folding and shearing, with slices of oceanic crust thrust northwards in the closing ocean basin.
(b) As the ocean basin continues to close, new thrusts are initiated behind the original thrust, which in turn has undergone inversion. The new thrust slices consist of a mixture of upper and lower oceanic crust to upper mantle (**serpentinite**).
(c) By ~370 Ma, deformation and ocean closure has resulted in the refolding of the initial thrust and stacking up of later thrusts onto the edge of the continental plate.
(d) Progressive overthrusting of the thrusts leads to the onset of small-scale partial melting and the eventual collapse of the thrust pile as it is finally emplaced onto the continental margin.
(From Jones, K. A. (1997) 'Deformation and emplacement of the Lizard Ophiolite complex...', *J. Geol. Soc.*, Vol 154.)

shortening are abundant and the grade of metamorphism is low, forming a weak cleavage. At some locations however (e.g. Figure 9.6c), higher degrees of deformation are apparent, forming localized thrusted folds and **duplex** systems. North of the Variscan Front, the level of crustal shortening is only sufficient to form gentle, open folds such as those in the unmetamorphosed Carboniferous rocks in County Clare, western Ireland (Figure 9.6d).

Figure 9.6 Examples of the decreasing intensity of folding northwards produced by the Variscan Orogeny across the southern British Isles.
(a) Chevron folds with recumbent axial planes, Millhook Haven, north Cornwall, equivalent to ~60% crustal shortening. Height of visible section ~15 m.
(b) Less intense, more irregular folds ~6 km north of (a) near Bude (SS(21)1907), produced by ~50% crustal shortening. Height of visible section ~13 m.
(c) Thrust cutting through folds in Carboniferous sandstone at Broad Haven (SM(12)8614), Dyfed, Wales. Height of section ~5 m.
(d) Gentle, open folding (~2–4% crustal shortening) on Mullagh More in the Burren, Co. Clare, Ireland.

The progressive change in shortening intensity and the effect this has had on the Older Cover strata of the south-western British Isles, is summarized in the two schematic cross-sections in Figure 9.7.

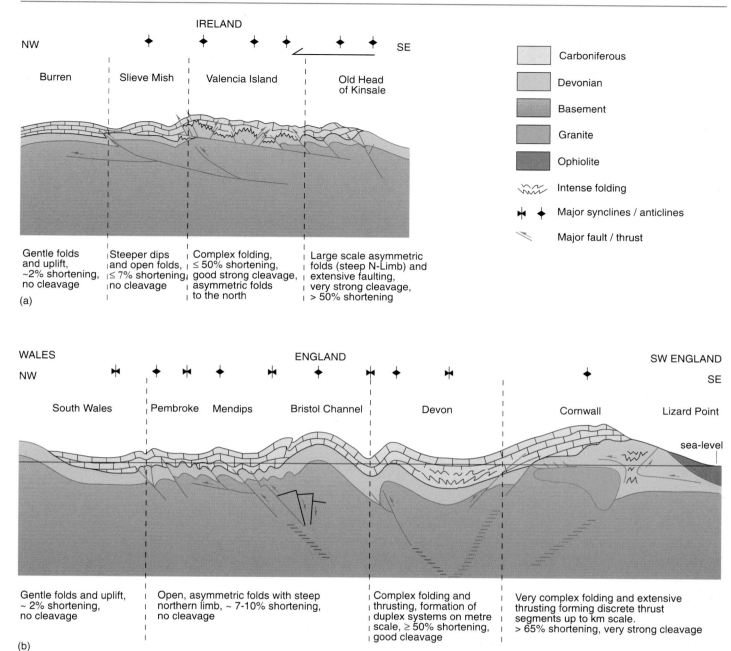

Figure 9.7 Schematic cross-sections illustrating the progressive increase southwards in folding, deformation and metamorphism associated with the Variscan Orogeny for (a) the south of Ireland from the Burren (County Clare) to Kinsale (County Cork) and (b) south Wales north of the Variscan Front to Lizard Point, south Cornwall.

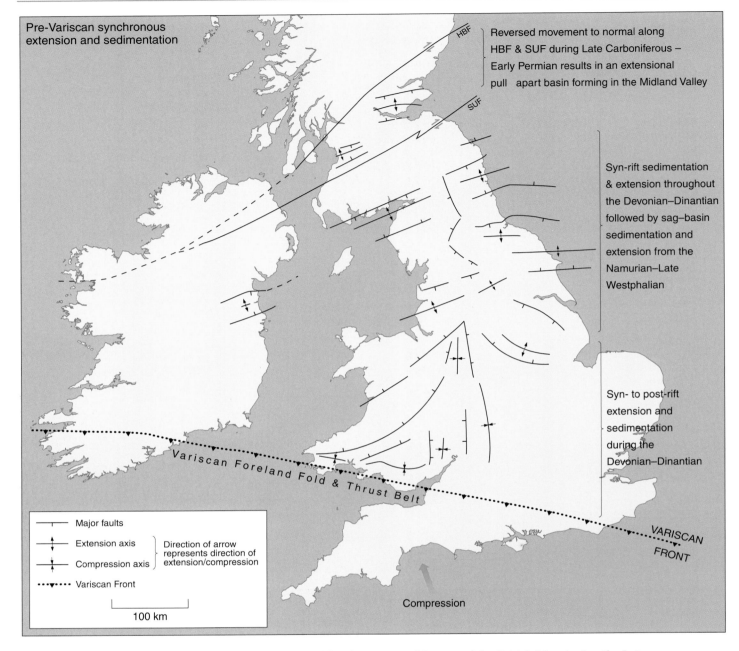

Figure 9.8 (a) Palaeogeographic map of the British Isles during the Late Carboniferous, illustrating major locations of (synchronous) syn-rift extension and post-rift (sag-basin) subsidence.

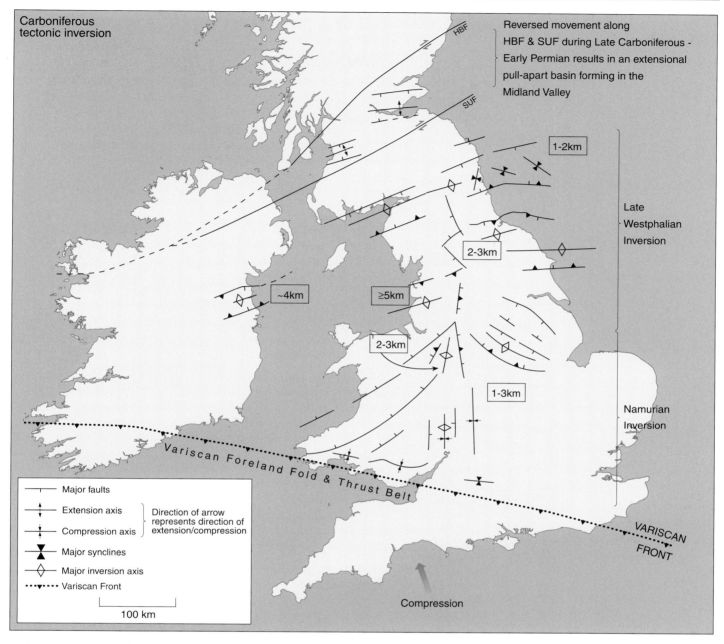

(b) Summary map of the amount of tectonic uplift, location and timing of tectonic inversion produced by the Variscan Orogeny. The boxed figures represent the approximate amount of uplift undergone by the sedimentary basins.
(From Peace, G.R. and Besley, B. M. (1997) 'End-Carboniferous fold-thrust structures, Oxfordshire, UK...', *J. Geol. Soc.*, Vol 154.)

9.2.5 Deformation of the Variscan Foreland

The Caledonian Orogeny resulted in crustal thickening and isostatic uplift to form a series of emergent highlands across the British Isles at the beginning of the Devonian. By the Early Carboniferous, rapid erosion rates had reduced many of these emergent blocks down to sea-level (Plate 6). Synchronous and subsequent thermal relaxation of the crust produced a series of extending and subsiding rifted troughs between the more buoyant blocks (Figure 8.8b), which continued to subside after extension had stopped, forming large-scale sag-basins (Figure 8.8c). With the onset of the Variscan Orogeny and closure of the Rheic Ocean, the tectonic regime across the British Isles changed from extensional (associated with crustal relaxation following the Caledonian Orogeny) to one of crustal shortening.

The unmetamorphosed region north of the Variscan Front is referred to as the Variscan Foreland. In this region, in addition to the strong E–W Variscan fold trends, some regional structures can still be seen to follow the NE–SW fold trend of the underlying Caledonian basement. This is particularly clear in the Bristol to Forest of Dean area (ST(31)57–SO(32)62), as well as further north-west between Brecon (SO(32)0528) and Builth Wells (SO(32)0450), and in the South Pennines near Bakewell (SK(43)2168), where the trend starts to swing round to the N–S. In these regions, two periods of folding are therefore evident, following the NE–SW Caledonian trend and the E–W Variscan trend.

The location and intensity of Variscan deformation across the Foreland has been strongly influenced by the block and trough morphology of the Older Cover and underlying basement. Below the blocks, the shallow post-orogenic granites and/or Caledonian basement have enabled these regions to remain unaffected by Variscan crustal shortening and deformation. In contrast, the sediment-filled troughs have undergone varying amounts of Variscan deformation, the amount and intensity of which is dependent on the orientation of each trough. E–W striking troughs have undergone the greatest amounts of shortening, whereas N–S trending troughs are affected to a much lesser extent. Similarly, E–W trending extensional (**normal**) faults underwent reverse reactivation (described below) due to shortening associated with the Variscan, whereas faults trending N–S were either unaffected or experienced some strike–slip movement.

Throughout the Late Devonian to Late Carboniferous, sediments poured into the troughs and the sag-basins in the Variscan Foreland. This synchronous sedimentation, extension and subsidence began progressively later northwards (Figure 9.8a). Variscan deformation did not reach north of the Cheviot Block, and so post-Caledonian extension and thermal subsidence remained dominant across northern England and the Midland Valley of Scotland, allowing the post-rift sag-basins to increase in volume.

From the Mid-Carboniferous, the effects of Variscan shortening were intensifying across the Foreland, with previously subsiding E–W trending basins undergoing significant amounts of shortening and uplift, while the movement direction along E–W normal (extensional) faults was reversed (Figure 9.8b). When the tectonic regime is reversed like this, it is referred to as **inversion tectonics**. The Variscan inversion resulted in many previously subsiding rifted troughs and sag-basins forming **pop-up basins** (Figure 9.9). Deformation and shortening associated with the orogeny continued until the end of the Late Carboniferous.

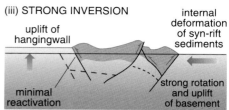
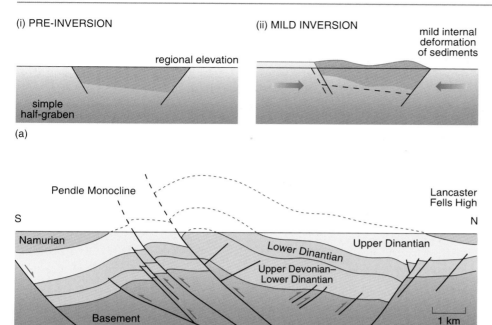

Figure 9.9 (a) Simplified model of how change from an extensional to compressional tectonic regime results in inversion and uplift of previously rifting basins:
(i) Pre-inversion graben, with both margins faulted. (ii) Mild inversion – shortening has resulted in minimal slip along the faulted margins, with the sediment in the trough gently folded. (iii) Strong inversion – the faulted margins have undergone reverse reactivation, while within the trough, the sediments have folded as well as faulted, due to high levels of shortening. (From Corfield, S. et al. (1996) 'Inversion tectonics of the Variscan foreland of the British Isles', J. Geol. Soc., Vol. 153.)
(b) Cross-section through the strongly inverted Bowland Basin, in the west of England.

9.3 Permo-Carboniferous igneous activity in the Variscan Foreland

In Section 9.2.1, we examined the post-orogenic granites in south-west England. In the Midland Valley of Scotland and the Northumberland Trough in northern England, a very different type of igneous activity occurred at the end of the Variscan Orogeny, associated with continued extension and passive continental rifting.

Extension and passive rifting resulted in the shallow-level intrusion of the extensive Midland Valley **Sill** and the Whin Sill (Figure 9.10a), which have both been radiometrically dated to ~295 Ma, making them Late Carboniferous in age. The episode as a whole is described as Permo-Carboniferous, because it straddles the Carboniferous/Permian boundary. The margin of the Whin Sill is not always parallel with the adjacent sedimentary units (Figure 9.10b). Systematic field studies have revealed that cross-cutting relationships also occur between the Midland Valley Sill and the adjacent Upper Carboniferous sediments. By investigating the thickness of both sills across their total outcrop, it has been shown that both intrusions are considerably thicker towards their centres than at their margins. These two lines of evidence indicate that the sills were intruded into concurrently subsiding sedimentary basins, rather than being intruded into the crust prior to crustal shortening and basin formation. If the sills had been intruded prior to shortening, they would have been expected to be relatively uniform in thickness. Instead, it is believed that both sills were fed by one or more feeder dykes at the edge of the sedimentary basins, with the magma supply flowing down between the sedimentary layers towards the lowest points, where it ponded.

Just south of Durham (NZ(45)2941), there is a series of basaltic dykes, trending approximately WSW–ENE. These are post-Westphalian but pre-Magnesian Limestone (Permian) in age.

Figure 9.10 (a) Map of Late Carboniferous (extensional) igneous activity across the Midland Valley of Scotland and northern England. (From Francis, E.H. (1982) *J. Geol. Soc.*, Vol. 139.)
(b) Contact relationship between the Whin Sill and adjacent Carboniferous sediments at Low Force, Middleton-in-Teesdale.

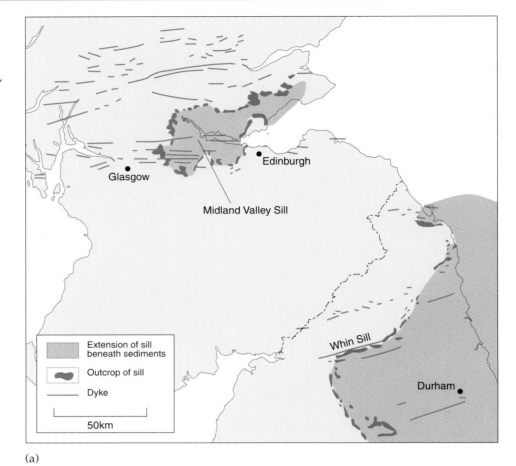

(a)

(b)

It is not known whether any of these dykes or similar examples further north (e.g. NY(35)4079–NZ(45)0097 and NY(35)5861–NY(35)9470) acted as the main feeder dyke for the Whin Sill. However, it is certain that they were all part of the same episode.

Not all Permo-Carboniferous igneous activity is intrusive; there are also extrusive rocks of Carboniferous and Permian age.

The Carboniferous extrusive units actually extend right through the Carboniferous. Over 80% of all activity occurred during the Early Carboniferous and consisted of rift-related mafic to felsic lava flows and pyroclastic eruptions (see Section 8.5, Figure 8.4). By the Late Carboniferous, although the rate and volume of volcanic activity had decreased in this region, the explosiveness had increased forming localized ash deposits from pyroclastic vents. Numerous examples of these can be found in West Lothian, Fife and the Clyde Valley.

Volcanism continued to decrease into the Early Permian, with only a few basalt lava flows found overlain by Permian sediments in the Southern Uplands (e.g. NX(25)89), the Midland Valley (e.g. NS(26)42) and in south-west England (e.g. SS(21)90). In contrast, more extensive Early Permian volcanic rocks occur beneath the North Sea and northern mainland Europe (Figure 9.11). These are related in part to the final stages of post-orogenic faulting as well as to continued lithospheric extension, basin subsidence and the initiation of rifting in the central North Sea. As you will see in Section 10, rifting stopped in the North Sea before sea-floor spreading could commence.

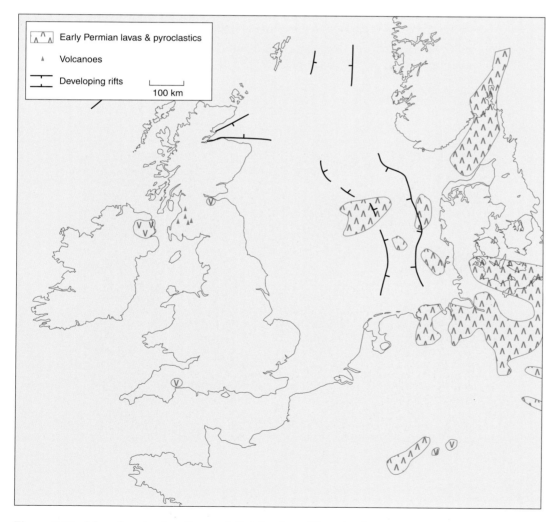

Figure 9.11 Map showing location and distribution of Permian igneous activity across the British Isles, North Sea and north-west European mainland

9.4 Summary

- The Variscan Orogeny was associated with the closure of the Rheic Ocean and collision of several Gondwana terranes, and then of Gondwana itself, with the ORS landmass.
- The main deformation zone was centred on southern to central Europe (now marked by the suture zone), and only the far south-west of the British Isles experienced strong deformation and metamorphism, the northern limit of which is represented by the Variscan Front.
- South of the Variscan Front, the Older Cover strata and Caledonian basement have undergone thrusting and complex folding accompanied by low (slate) grade metamorphism. Variscan deformation is characterized by asymmetric folds with predominantly E–W trending fold axes and steepened fold limbs on the hangingwall side.
- North of the Variscan Front, the Older Cover and basement are moderately to slightly deformed but unmetamorphosed, with inversion tectonics leading to the formation of pop-up basins and the reverse re-activation of E–W extensional (normal) faults.
- North of the Alston Block, post-Caledonian extension and thermal subsidence continued as the predominant processes, with passive continental rifting resulting in the emplacement of shallow-level intrusives and the localized eruption of explosive mafic–felsic pyroclastics.
- By the Permian, this volcanism was beginning to decrease, with the remaining activity focused in the central North Sea as it attempted to rift apart.

10 THE YOUNGER COVER

10.1 INTRODUCTION – FROM DESERTS TO GLACIERS

By the end of the Variscan Orogeny, the assemblage of nine basement terranes that make up the British Isles had been completed (Figure 5.2 and Table 5.1). One of the outcomes of this orogeny and associated basement uplift was the renewal of erosion and sedimentation processes that led to a regional unconformity followed by onset of deposition of the Younger Cover, which represents the final phase of lithotectonic development of the British Isles.

The Younger Cover consists of Permian to Pleistocene (the lower part of the Quaternary) strata (see Appendix). It includes sandstone, breccia, mudstone, dolomite and evaporite deposits in the Permian and Triassic, and organic-rich mudstones, sandstones, **marls** and limestones in the Jurassic to Cretaceous. The Tertiary and Pleistocene are mostly represented by mudstones and sandstones.

❑ Describe the age of the *base* of the Younger Cover across the British Isles. (Refer back to Section 5.6 if necessary.)

■ Across the British Isles, the base of the Younger Cover varies from Permian to Triassic age.

For example, in the Vale of Eden (NY(35)45–81) the base of the Younger Cover consists of Permian, whereas from Cardiff to Newport (ST(31)27–49) the Permian is missing, so the base is Triassic in age. In virtually all locations across the British Isles, the base of the Younger Cover (regardless of its age) lies unconformably on top of older rocks. This unconformity is in places irregular, and represents a buried palaeotopography.

As discussed in the previous Section, collision between the ORS landmass (containing the British Isles) and Gondwana during the Carboniferous, culminated in the formation of the supercontinent Pangea (Figure 3.1e–f). At this time, the British Isles occupied an intracontinental setting, and was located just north of the equator at ~10°–20° N. Over the next 290 million years, the British Isles drifted slowly northwards with the supercontinent, which began to break apart in the Jurassic (Figure 3.1g), reaching its final latitude of ~51°–58° N by the Mid-Cenozoic Era (Figure 3.1h). This northwards drift in association with the opening of the Atlantic and the establishment of a warm ocean current (the Gulf Stream), once the Atlantic was wide enough, had a significant effect on the climate of the British Isles, causing it to vary from hot desert conditions to semi-arid, temperate and glacial conditions.

In this Section, you will read about the Mesozoic and Cenozoic evolution of the British Isles, excluding the last 2 million years. In particular, the interaction between climatic effects, sea-level changes, tectonic movements and igneous activity will be examined in terms of their influence on the type and extent of strata that were deposited. The final 2 million years of the geological history of the British Isles that make up the Quaternary will be examined in Section 11.

10.2 CHANGING SEA-LEVELS FROM THE PERMIAN TO CRETACEOUS

Question 10.1 Examine Figures 3.1–3.2 and describe how the global sea-level and distribution of continental crust varied between 290 and 85 Ma (Permian to Late Cretaceous Periods).

These changes in global sea-level together with the break-up of Pangea had a significant effect on the types of depositional environment that occurred across the British Isles during the Mesozoic and Cenozoic Eras.

Question 10.2 During the Carboniferous, what were the predominant depositional environments over a significant proportion of the British Isles? What rock types formed in these areas? (Refer back to Section 8.6.2 if you cannot remember.)

These Carboniferous rocks were deposited when the British Isles was at ~0°–15° N, located on the edge of a back-arc basin in the ORS landmass. Similar depositional environments (though in different plate tectonic settings) can be found today in the Caribbean (shallow-water marine limestones) and Nigeria (deltaic successions).

During the Permo-Triassic, the British Isles occupied an intra-continental site (within Pangea), drifting northwards to ~10°–20° N, equivalent in latitude to the present day Saharan desert (Figure 3.1f). Throughout this period, erosion of the recently uplifted landmass and increased sedimentation rates formed the New Red Sandstone. As with the ORS, the New Red Sandstone consists of red aeolian sandstones (Figure 10.1a), together with alluvial, fluvial and lake deposits (Figure 10.1b), which are interbedded with some shallow-water marine and evaporite deposits (Plates 10–11). These are indicative of an arid, desert environment that experienced periodic incursions by a shallow, ephemeral land-locked sea, within which the water had a restricted circulation. A range of Permian lithologies is seen on either side of the Pennines.

As the global sea-level began to rise during the Early Jurassic, the desert conditions and land-locked transient seas of the Permo-Triassic were replaced by more permanent warm shallow-water shelf seas, within which the seawater was able to circulate more freely. Detailed palaegeographic studies reveal that the Mesozoic global sea-level rise occurred in a series of pulses, as evident on Figure 3.2, replacing shallow-water shelf, coastal and continental sedimentary environments by environments that were characterized by lower-energy (deeper-water) marine deposits.

❑ What changes in the sedimentary successions would be seen as these marine transgressions occurred?

■ The shallow-water coarse-grained siliciclastics and limestones would be replaced by deeper-water siltstones, mudstones and limestones.

Figure 10.1 (a) Permian aeolian cross-stratified sandstones (sand dunes) at Bongate Scar near Appleby (NY(35)6721). Note the scale of the cross-stratification at this exposure. (b) Permian **wadi** deposits at Hoff Quarry, near Appleby. The student is examining a lens-shaped section through a small (~1.5 m) channel. (The term 'wadi' is used to describe a valley with an intermittent stream, that develops in semi-arid or arid conditions.)

(a)

(b)

Throughout the remainder of the Mesozoic Era, the observed variations in sediment types can be explained in terms of a long-term progressive increase in global sea-level. In the Early Jurassic (206–180 Ma), thick successions of deep-water limestones and mudstones were deposited across the Midlands and south-west England (Figure 10.2a). These grade up into a repeating sandstone, mudstone and limestone succession across Yorkshire and the North Sea regions (Figure 10.2b, Plate 12). By the Mid-Jurassic (~180–158 Ma), the global sea-level began to stabilize, permitting the formation of shallow-water **oolitic limestones** in the southern British Isles which graded up into deltaic and lake sediments in Lincolnshire, Yorkshire and the North Sea (Plate 13). At the start of the Late Jurassic, the global sea-level began to rise once more, deepening the sedimentary basins and allowing deposition of the Oxford and Kimmeridge Clays (Figure 10.2c) across the whole of the submerged area of the British Isles (Plate 14). The Jurassic succession is discussed more fully in Section 10.5.2.

Figure 10.2 (a) A ~25 m thick repeating succession of Lower Lias carbonate shales from the north Somerset coast. In the centre of the photograph, an extensional (normal) fault can be seen cutting through the strata.
(b) Middle Lias shallow-water marine sandstones, with excellent **bifurcating** ripple marks on the sedimentary palaeosurface, Staithes, North Yorkshire.
(c) Thick repeated succession of Upper Jurassic clays, Kimmeridge, Dorset (SY(30)9278), deposited under quiet deep-water marine conditions.

(a)

(b)

(c)

During the Early Cretaceous, tectonic movements overshadowed the effect of global sea-level changes in the British Isles, with the deposition of deep-water carbonate clays followed by shallower-water **greensands*** in more marginal areas (Plate 15) even though globally the sea-level was rising. By the Late Cretaceous, the rise in global sea-level dominated the range and location of sediments once again, with the formation of thick successions of **chalk** (a limestone described in Section 10.5.3) across virtually all of the British Isles except for a few remaining upland areas (Plate 16).

Therefore, although there is evidence for periodic falls in sea-level, the overall signature throughout the Jurassic and Cretaceous Periods was one of a marine transgression that swept across the lowlands of the British Isles (Plates 12–16), deepening the sedimentary basins.

10.2.1 Mechanisms of sea-level change

Question 10.3 What mechanisms could have influenced the global sea-level changes that occurred from the Permian to Cretaceous described above? (You may find it useful to refer back to Section 3.)

Sea-level changes produced by glacial mechanisms can occur at a rate of up to 1 cm per year, whereas tectonically controlled sea-level changes are much slower, occurring at ~1 cm per thousand years. The rate at which sea-level change occurs as a result of lithospheric extension is wholly dependent on the extension rate, but generally falls between the rates of the previous two mechanisms.

So how do these three mechanisms fit with the observed changes in eustatic sea-level from the Permian to the Cretaceous? The fall in sea-level that started during the Late Carboniferous (Figure 3.2) can be partially attributed to the glaciation of Gondwana. Furthermore, the closure of the Rheic Ocean during the Variscan Orogeny and the assembly of Pangea left few active mid-oceanic ridges during the Permian (Figure 3.1e), displacing relatively little water onto the continental shelves.

Lithological and faunal changes over the Triassic–Jurassic boundary can be used to infer an average rate of sea-level rise that is too fast for tectonic–eustatic processes. Unlike the Late Carboniferous however, the lack of any significant glacial ice and the warm global climate at this time (Figure 3.2), means sea-level changes cannot be attributed to glacial–eustatic processes. Recent work on the tectonic evolution of a number of key Triassic–Jurassic locations around the globe (including the Mediterranean and North Sea), has shown that many of these areas were characterized by extensional tectonism associated with the extrusion of significant amounts of basaltic material and continental rifting, prior to ocean basin development (see Section 4.2.1). Although these localized extensional rift-basins produced regional epeirogenic rather than true eustatic sea-level changes, this occurred over extensive geographical areas rather than being confined to isolated basins. For example, during much of the Jurassic, extension in the present day North Sea area affected sea-levels over vast areas of north-western Europe including the British Isles. At the same time, in southern Europe and North Africa, sea-level changes were associated with extension in the Tethys (proto-Mediterranean) region (Figure 3.1f–g).

In some instances, geologically instantaneous falls in global sea-level can be produced by the release of lithospheric tension associated with extension. As an extensional basin finally rifts to form an ocean ridge, the associated release of tension allows the lithosphere to 'spring-back' (like a snapped elastic band), producing a rapid fall in sea-level of ~1 cm per year, comparable to glacio-eustatic rates. Examples of this process have been found across north-western Europe.

---

* Greensands are sandstones that contain a high abundance of the mineral **glauconite**, giving fresh rock surfaces a bright green colour. Glauconite is a sheet silicate mineral of the mica group and is a common constituent of marine sediments, forming during or soon after the sediments are deposited (i.e. **diagenetically**).

THE YOUNGER COVER

From the Mid-Jurassic (~180 Ma) to the Early Cretaceous, the rate of sea-level rise decreased to an average of ~1 cm per thousand years, which can be attributed to climate change and to tectonic–eustatic processes on a global scale, associated with the rifting and break-up of Pangea (Figure 3.1f). This rifting produced a new network of constructive plate boundaries responsible for forming the present day Atlantic, Indian and Southern Oceans. Therefore in answer to Question 10.3, the Permian to Cretaceous global sea-level changes can be attributed to a combination of glacial–eustatic, tectonic–eustatic and lithospheric extensional processes.

10.3 MESOZOIC TO TERTIARY IGNEOUS ACTIVITY

After the Permo-Carboniferous igneous activity (Section 9.3), the next major phase of igneous activity in the British Isles did not occur until the Mid-Jurassic (Figure 10.3).

10.3.1 RIFTING IN THE NORTH SEA

From the Mid-Triassic to Mid-Jurassic, the centre of the present-day North Sea started to dome up, forming a structural high that was episodically emergent (Plates 11–13). This dome (referred to as the Mid North Sea High), divided the North Sea into two basins, and had a significant impact on the development of the petroleum reserves in the North Sea. As the Mid North Sea High increased in size, the Anglo–Brabant landmass prevented the basin on the south side of the High from extending southwards, resulting in compression and widespread mobilization of the Permian evaporite successions (which flowed under

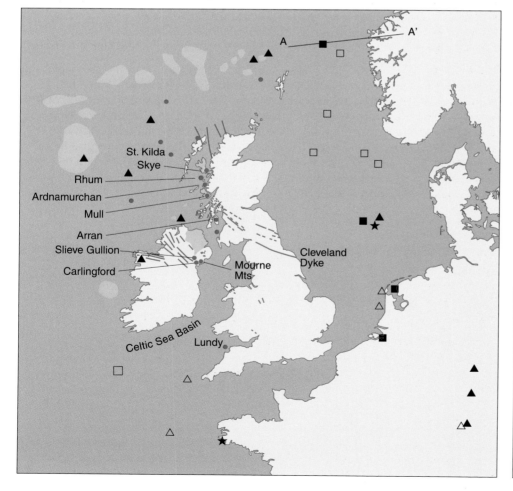

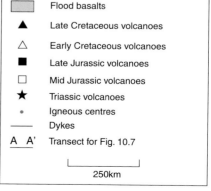

Figure 10.3 Map of the location and distribution of igneous activity in the British Isles, north-west Europe and surrounding ocean basins during the Mesozoic (Triassic–Cretaceous) and Tertiary. (The transect A–A' relates to Figure 10.7.)

pressure like magma). This produced a series of **salt domes** and pillars that deformed and cut through the overlying strata. Meanwhile in the northern basin, which was not structurally confined, continuous rifting and subsidence occurred in the Viking, Moray Firth and Central Grabens (Plate 12). This led to thick sedimentary successions accumulating in the grabens, while the lithosphere continued to extend, dome up and undergo rapid erosion. By the Mid-Jurassic, extension was sufficiently great to permit the intrusion of tholeiitic to alkali sills and dykes and the eruption of alkali basalts, ash and other pyroclastic material, at the junction between the Moray Firth, Viking and Central Grabens (Plate 13) which was a point of weakness. The igneous activity found to the south-west of the Celtic Sea Basin (Figure 10.3), can also be attributed to lithospheric extension and rifting in the Celtic Sea and Western Approaches grabens.

From the Late Jurassic onwards, the rate of extension began to decline, which in turn impacted on the extent of volcanic activity. By the Early Cretaceous, extrusive activity was limited to two narrow rift zones in the far south of the North Sea, extending down the Central Graben into mainland Europe (Figure 10.3), and to the south-west of Land's End, at Wolf Rock (SW(10)2512).

10.3.2 Opening of the North Atlantic

During the Late Cretaceous and Early Tertiary, the hub of igneous activity switched from the North Sea rifting zone over to the west of the British Isles (Figure 10.3 and Plate 17). Here a N–S zone of mantle plumes resulted in active continental rifting separating the British Isles and Scandinavia from Greenland, and opening the North Atlantic Ocean. The entire length of the Atlantic Ocean did not, however, open all at once. The southern Atlantic (between South America and Africa) began to open in the Jurassic, followed by the central Atlantic (between North America and Africa) which opened in the Early Cretaceous. This was then followed by the North Atlantic which did not begin to open until the Late Cretaceous.

Within the British Isles and adjacent continental shelves, although the bulk of Early Tertiary igneous activity associated with this rifting event occurred over a relatively short period (63–51 Ma), the volume of material erupted was huge. Over 12 million years, a 2 km thick sequence of tholeiitic flood basaltic lava flows was built up. These flood basalts covered an area that was originally 1×10^6 km^2, extending across the north-west of Scotland and the north of Ireland (e.g. the Skye and Antrim flood basalts), as well as eastern Greenland as the basin started to rift (Figure 10.3 and Plate 17).

In addition to this extensive extrusive activity, this period was also characterized by the formation of numerous shallow-level minor intrusions including cone sheets, ring dykes and **radial dykes**, trending WNW–ESE across the northern British Isles (Figure 10.3). These include the Cleveland Dyke, believed to originate from Mull. It runs across northern England entering the North Sea basin just north of Scarborough (SE(44)8999) – its last outcrop on land.

10.4 Tectonic development of the Younger Cover

10.4.1 Development of Permo-Mesozoic half-grabens, basins and highs

Question 10.4 Look at the palaeogeographic map in the Appendix and describe how the thickness of the Younger Cover deposits varies across the British Isles and continental shelf basins.

An insight into this distribution can be gained by examining the Permo-Triassic rocks in the Vale of Eden between south Carlisle, Penrith and Kirkby Stephen (NY(35)4552–5131–7806). Close examination of the Vale of Eden reveals that its western margin is an unconformity, with basal Permian sandstones and breccias resting on top of the Lower and Upper Carboniferous. In contrast, the eastern margin is faulted, with Upper Permo-Triassic sandstone juxtaposed against Carboniferous to Ordovician strata.

This is an example of a half-graben, controlled by extensional (normal) faulting that occurred synchronously with sedimentation (i.e. **syn-rift sedimentation**). This faulting increased the depth of the basin at its eastern edge, while the western unconformable margin remained relatively fixed, causing the basin floor to be progressively tilted to the east (Figure 10.4).

Other examples of extensional half-grabens can be found across the British Isles and surrounding continental shelves. For example, along the coastline of Cardigan Bay between Tywyn (SH(23)5900) and Harlech (SH(23)5830), two small outcrops of Oligocene can be seen to be faulted against Lower Palaeozoic rocks, with **downthrow** to the west. Geophysical and borehole evidence shows that the Oligocene is underlain by a thick succession of Lower Jurassic and Triassic rocks that thin towards the west as a result of syn-rift sedimentation (Figure 10.5).

At its broadest level, this model can be used to explain the changing thickness of Younger Cover across the whole of the British Isles, with the thickest successions forming on the downthrown side of a half-graben, which underwent syn-rift sedimentation. As you will soon see, superimposed on this model is a series of smaller structural features.

The outcrop width of the Jurassic between Lincoln (SK(43)9872) and Market Weighton (SE(44)8842) on the eastern side of the British Isles becomes narrower towards Market Weighton. It is not possible to determine whether this narrowing is due to an increase in the easterly dip of the strata nearer to Market Weighton or whether it is because the beds thin to the north. Field information reveals the latter explanation to be correct, indicating that the southern area, where the strata are thicker, must have subsided more rapidly than the north.

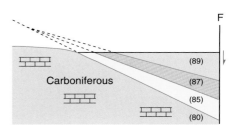

Figure 10.4 Schematic sketch showing how the Vale of Eden was progressively faulted and filled during the Mesozoic, to produce a half-graben structure.

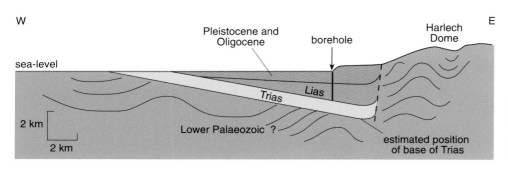

Figure 10.5 Sketch cross-section based on geophysical and borehole data of the Younger Cover resting unconformably on Caledonian basement in Cardigan Bay.

This thickening and thinning pattern is a common feature of the Mesozoic in the British Isles. The thin, condensed successions formed on discrete higher areas of bedrock or 'highs' (some of which formed small islands at the edge of a Mesozoic landmass), whereas the thicker expanded successions were deposited on the surrounding slopes and intervening basins (Figure 10.6).

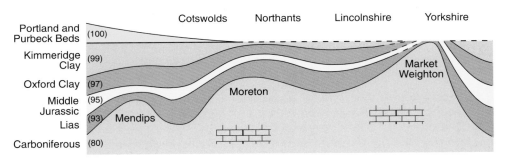

Figure 10.6 Sketch cross-section of the Mesozoic basin and high topography between Dorset and Yorkshire. The strata are thickest in the basins and thinnest over the highs.

As well as the Market Weighton High, which formed a positive feature throughout the whole of the Jurassic, two other significant highs influenced the location and extent of Mesozoic sedimentation in the southern British Isles. These highs occurred in the Mendips (ST(31)6320–7590), and the Vale of Moreton, which was located at the western end of the Anglo–Brabant landmass (Plates 12–14).

The cause of the basin and high topography can be related to differential vertical movements associated with uplift of the buoyant highs and subsidence of the surrounding basins. This was produced by reactivation of Variscan and Caledonian fault structures in the basement, associated with Permian to Early Cretaceous rifting in the North Sea. In addition during the Jurassic, the opening of the central Atlantic Ocean to the south-west of the British Isles may also have contributed to these vertical movements.

Based on variations in the preserved stratigraphic succession, some highs (e.g. the Mendips) are known to have periodically subsided at a similar rate to the adjacent basins, before undergoing a later phase of uplift. It is unclear why the highs formed buoyant structures, but two hypotheses include the presence of low-density crustal rocks at their cores, or some structural influence related to the location of Variscan fold axes. There is indeed some geophysical evidence for an underlying granite at the site of the Market-Weighton High.

10.4.2 Permo-Mesozoic extensional tectonics

The progressive development and deepening of the Permo-Mesozoic basins across the British Isles can be attributed to episodic E–W rifting of the North Sea and North Atlantic, and to interplay between post-orogenic extension and thermal subsidence associated with the Variscan Orogeny and Late Carboniferous igneous activity. Over the last few decades, petroleum exploration has revealed the complexity of the structural geology of the basins.

In general, the Permo-Mesozoic basins on land in the south of the British Isles have a predominantly E–W strike, associated with N–S extension along pre-existing E–W trending Variscan faults. In contrast, in the Midlands and north of England, where the effects of the Variscan Orogeny were much weaker, the Permo-Mesozoic basins have an approximate N–S strike. This is associated with the E–W extension controlled by structural trends in the Caledonian basement, and is approximately parallel to rifting trends to the east (in the North Sea) and west (in the North Atlantic). Some faults of this kind are seen in the cliffs near Staithes, North Yorkshire (NZ(45)7818).

In present-day offshore regions, passive continental rifting and NE–SW extension were predominant in the Late Carboniferous to Early Permian, occurring in response to the post-Caledonian thermal subsidence that had been ongoing since the Devonian. During the Late Permian, as the orogenically thickened lithosphere continued to cool, subside and undergo erosion, isostatic readjustment led to extension and the formation of the intracontinental North Sea sedimentary basin. As extension proceeded, a series of small extension rift-basins and half-grabens formed across the basin (Figure 10.7).

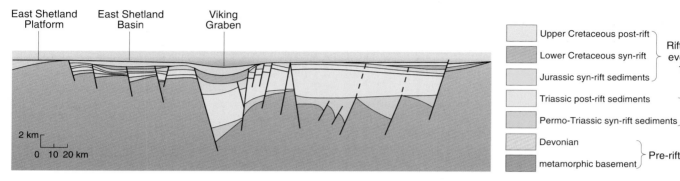

Figure 10.7 Sketch cross-section across the Viking Graben in the North Sea. (The location of this basin is shown on Plates 11–15.) Note that the Mid-Cretaceous age of faulting can be seen clearly as these structures extend through the Lower Cretaceous but not the Upper Cretaceous sediments. Some of these fault structures form important traps for oil in the North Sea. (From Færseth, R. B. (1996) 'Interaction of Permo-Triassic and Jurassic extensional fault-blocks...', *J. Geol. Soc.*, Vol. 153.)

During the Mid-Triassic to Mid-Jurassic, thermal subsidence took over as the primary basin-forming process in the North Sea region, culminating in the formation of large sag-basins across the tops of the Permo-Triassic rift-basins (e.g. see Figure 8.8). However, rifting was renewed in the northern North Sea basin from the Mid-Jurassic to Early Cretaceous (Figure 10.7). Associated with this second phase of rifting was the intrusion of a number of tholeiitic to alkali sills and dykes, along with the eruption of alkali basaltic lavas in the northern North Sea. This igneous activity occurred synchronously with rapid sedimentation and continued rifting. Evidence for syn-rift sedimentation can be found in all onshore and offshore basins of Permian to Early Cretaceous age.

Overall, extension during the Mid-Jurassic to Early Cretaceous had a profound effect on the structure of the present-day continental shelf around the British Isles, forming the Viking, Moray Firth, Central and Southern Grabens in the North Sea, and the Western Approaches, Celtic Sea and Bristol Channel basins in the south-west. Therefore to summarize, unlike the Older Cover structures that were controlled by compression and inversion tectonics (see Section 9.2), the predominant Younger Cover structures up to the Early Cretaceous were produced by extensional tectonics, related to rifting in the North Sea and North Atlantic.

10.4.3 Tertiary inversion tectonics and the Alpine Orogeny

By the Late Cretaceous, although sedimentation was continuing in the offshore basins due to thermal relaxation, virtually all extensional faulting had ceased (Figure 10.7), allowing sediments of a more uniform thickness to be deposited across the whole basin. At the same time as rifting stopped, the south-east of England began to rise above sea-level, forming the shallow Weald Dome (~200 km long by up to 80 km wide), extending across Kent to Hampshire. This marked the start of a period of compression and the re-instigation of inversion tectonics across the British Isles, which continued throughout the Tertiary.

The cause of this compression can be related to two events. First, at the end of the Cretaceous, the northward-drifting African Plate collided with southern Europe. This caused the Alpine Orogeny with the main focus of deformation occurring in the present day Mediterranean. Secondly, to the west of the British Isles, the release of lithospheric tension due to spring-back resulted in compression and uplift as the southern extremities of the North Atlantic Ocean changed from a rifting to ocean spreading environment (see the end of Section 10.2.1).

Throughout the Late Cretaceous to Tertiary, deformation in the British Isles was much gentler than during the previous two orogenies. In fact, the final amount of shortening was less than the amount of extension that had occurred in the British Isles during the Mesozoic. This resulted in a series of gentle deformation structures forming (e.g. E–W trending low-angle thrusts, **buckle folds** and **monoclines**, which are one-limbed folds, Figure 10.8), that were produced by the reactivation of pre-existing structural weaknesses in the crust (e.g. the Purbeck–Isle of Wight Monocline, Dorset SY(30)8080–SZ(40)6485, and the Chalk Hogs Back, Guildford TQ(51)0050).

Figure 10.8 Example of a small monoclinal fold in the field. This example is actually in the Lower Carboniferous strata south of Castletown on the Isle of Man, but is similar to the larger-scale monoclines found across the south of the British Isles.

The sedimentary and structural history of the Younger Cover can therefore be summarized in three main steps (Figure 10.9):

(a) *Permian–Early Cretaceous*: sedimentary deposition in a series of tilted half-grabens formed by extensional faulting above reactivated Variscan and/or Caledonian reverse faults (Figure 10.9a);

(b) *Late Cretaceous*: extensional faulting stopped, allowing Upper Cretaceous sediments to be deposited more uniformly across the basins (Figure 10.9b);

(c) *Tertiary*: reverse reactivation of the Mesozoic extensional faults, forming a series of low-angle thrusts, buckle and monocline folds (Figure 10.9c–d), associated with low levels of compression.

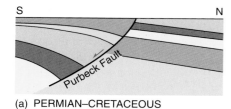
(a) PERMIAN–CRETACEOUS

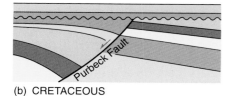
(b) CRETACEOUS

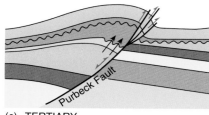
(c) TERTIARY

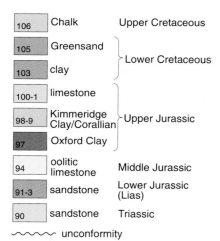

Figure 10.9 Schematic cartoons illustrating the evolution of the Lulworth region, Dorset during its syn-rift, post-rift and tectonic inversion history.
(a) Permian–Early Cretaceous syn-rift deposition: sedimentation into the graben and normal faulting along the Purbeck Fault occur contemporaneously, to produce strata that increase in thickness towards the fault plane.
(b) Late Cretaceous post-rift deposition: significant fault movement stops, allowing the Upper Cretaceous Chalk and Greensand to be deposited in uniformly thick layers across the underlying Jurassic sediments and structures.
(c) Tertiary tectonic inversion: compressional tectonics to the south result in reverse reactivation of the Purbeck Fault, producing a faulted monocline. (From Underhill, J. R. and Paterson, S. (1998) 'Genesis of tectonic inversion structures...', *J. Geol. Soc.*, Vol. 155.)
(d) The Lulworth Crumple, which formed in response to flexural slip in the hangingwall between the Portland and Purbeck beds and the Wealden Group (SY(30)8379).

(d)

10.5 Palaeogeography of the Younger Cover

The following discussion uses the palaeogeographic maps (in the Plates section in the centre of this book) to illustrate the influence of changing sea-levels, igneous activity, tectonism and palaeolatitude (and hence climate), on the geological development of the Younger Cover.

10.5.1 Permian and Triassic (10°–20° N, 290–206 Ma)

Permian

A marked unconformity separates the basal Permian sandstones and breccias from the Older Cover and basement, which formed a mountainous and arid desert environment at this time. In the British Isles, the Lower Permian consists of a mixture of breccias and poorly-sorted sandstones that were deposited by flash floods as alluvial fans (Figure 10.1b). It also consists of sandstones with large-scale cross-stratification (Figure 10.1a) that were deposited in aeolian sand dunes between the edge of the mountainous regions and the margin of the intracontinental Zechstein Sea. The high **porosity** and **permeability** of these aeolian sandstones (which are called the Rotliegende or 'red layers' in Europe) have made them economically important, as they form the main gas reservoirs in the southern North Sea.

Throughout the Permian, a combination of lithospheric uplift and fluctuations in sea-level periodically cut the Zechstein Sea off from the open marine Tethys Ocean, which was situated across south-east Europe. This prevented the Zechstein from circulating freely which, combined with arid conditions, permitted prolonged periods of evaporation and desiccation to form substantial evaporite successions. The type of evaporite deposits that formed under these conditions is directly linked to the salinity of the seawater. Once evaporation has reduced the original volume of seawater in any restricted basin to ~50%, the salinity of the remaining water is sufficiently high to allow evaporite salts to precipitate out of solution. Continued evaporation and precipitation of salts leads to the formation of an evaporitic succession. The first minerals to precipitate out of solution in this sequence are the carbonates (calcite, $CaCO_3$ and dolomite, $CaMg(CO_3)_2$), which are followed by the sulphates (**gypsum**, $CaSO_4.2H_2O$ or **anhydrite**, $CaSO_4$). At very high evaporation and salinity levels, chloride salts form, starting with **halite** (NaCl) followed by the potassium (KCl) and magnesium salts ($MgCl_2$).

During the Late Permian, although Scotland and north-east Ireland were still covered by aeolian sand dunes, the environment of deposition changed across the present day North Sea and parts of eastern England, as the Zechstein Sea engulfed this area (Plate 10). Prolonged periods of evaporation combined with episodic replenishment of the Zechstein Sea resulted in a thick succession of dolomite (known as the Magnesian Limestone) forming across north-east England and the western margin of the North Sea. These dolomites are intercalated with thinner bands of marl (calcareous mudstones), which represent the start of each new replenishment episode. Farther out in the basin, substantial sulphate and salt deposits developed. These thicken eastwards and form the cap or trap-rock across many of the gas and oil reservoirs in the North Sea.

In the Midlands and south-east England, the Upper Permian is either absent or represented by red marls. Towards the end of the Late Permian, uplift across the centre of the North Sea started to form a barrier, which continued to grow throughout the Triassic to Jurassic Period.

Triassic

The start of the Triassic is marked by a marine regression across a large proportion of north-west Europe. This is associated with the global fall in sea-level caused by the uniting of the continents in Pangea (Section 10.2 and Question 10.1). In the British Isles, where the Permian and Triassic strata are found in contact, a continuous succession occurs. Elsewhere in the British Isles, the Triassic sits unconformably on top of Older Cover or basement.

Lower Triassic sediments are found in north-east Ireland, south-east Scotland and north-west England as well as across much of Europe. They are dominated by conglomerates and pebbly sandstones (Plate 11), deposited by **braided** rivers. These flowed across the continental terrain, sourced from higher lands in Brittany and the lower lying Anglo–Brabant and Welsh landmasses. During the Late Triassic, local extension outpaced the still falling global sea-level to produce a local epeirogenic rise in sea-level across much of England, Wales and eastern Ireland. Seawater from the Tethys Ocean flooded the area, but only shallowly, leading to the deposition of evaporite successions comprising anhydrite ($CaSO_4$) and gypsum ($CaSO_4.2H_2O$), with rarer examples of halite ($NaCl$) also forming in some areas.

10.5.2 Jurassic (30°–40° N, 206–142 Ma)

Lower Jurassic

During the Early Jurassic, global sea-level was rising and the continental Permo-Triassic desert sediments were replaced by a cyclic succession of fossiliferous dark grey marine mudstones, marls and limestones (Figure 10.2a). This marked the beginning of marine conditions across the low-lying areas of the British Isles (Plate 12). The mudstones are organic-rich and form an economically important source of oil in parts of the British Isles (e.g. oilfields along the Dorset coast), but are not the source of the North Sea oilfields, which are mainly Upper Jurassic (Figure 8.12).

In addition to the limestones and marls that formed under low-energy conditions, condensed successions of oolitic limestones were deposited directly on top of the Palaeozoic basement in some areas such as the Mendips and Glamorgan. These condensed successions are typical of the marginal areas of the basin topography found throughout the Mesozoic (Figure 10.6). The thinner layers of shallow-water facies indicate the location of the highs, which are flanked by thicker deposits of deeper-water facies that formed in the basins.

In the north of England and west of Scotland, the Lower Jurassic cyclic succession becomes coarser-grained, and consists of alternating layers of sandstones and siltstones, indicative of shallower water and a higher energy of deposition. On a local scale, these sandstones and siltstones are frequently capped by shelly limestones and **ironstones**, which were economically important sources of iron in the British Isles until the 1940s.

Ironstones are clay-rich sedimentary rocks that are generally nodular in form, and characterized by a high abundance of iron. This iron is typically present in the form of **berthierine** (formerly known as chamosite), a sheet silicate mineral similar to chlorite, with the approximate formula $Fe_5Al_2Si_3O_{10}(OH)_8$, **siderite** ($FeCO_3$) or **goethite** ($FeO.OH$), which occur as fine grains in the sedimentary matrix or as discrete ooids. In order for ooids made of these minerals to form, the sedimentation rate must be low. As in the case of more familiar calcium carbonate ooids, warm shallow seawater is required, with wave action necessary to wash the grains back and forth. Some of the best examples of ironstones in the British Isles can be found in Northamptonshire (in the East Midlands of England), Raasay (off the east coast of Skye) and Cleveland (on the north-east coast of England).

Mid-Jurassic

During the Mid-Jurassic, uplift associated with extension and rifting produced an emergent area of land (the Mid North Sea High), that extended across both the North Sea and a large proportion of northern England and Scotland (Plate 13). Erosion of this highland resulted in the deposition of fluvial and deltaic sediments around its margins, interbedded with marine strata. The episodic fluctuation between continental and marine sediments may be related to periods of sudden uplift accompanied by rifting in the North Sea, resulting in the deposition of continental sediments, subsequently followed by subsidence and a marine transgression.

At this time, rifting in the North Sea was accompanied by the eruption of a series of alkali basaltic lavas and pyroclastic material, each flow being up to 9 m thick (Figure 10.3 and Plate 13). These basaltic lavas are not the first indication of volcanism in the Mid-Jurassic, because slightly older layers of fine-grained volcanic ash are found across the Midlands of England and in Skye. Similar ash layers are also found across the Bath region in south-west England, originating from eruptions at the southern end of the Celtic Sea Basin rather than the northern North Sea (Plate 13).

Across the submerged southern regions of England, thick successions of oolitic limestones (e.g. the Inferior and Great Oolites) were deposited. These units indicate that at this time, the southern regions of England were covered by a shallow warm sea that was not subjected to significant amounts of siliciclastic deposition. At the end of the Mid-Jurassic, these shallow-water shelf regions were replaced by deeper-water shelves and basins.

Late Jurassic

By the Late Jurassic, rifting was beginning to wane in the northern North Sea, causing the Mid North Sea High to subside progressively below sea-level, as isostatic re-equilibration of the crust occurred (Plate 14). This produced a rise in the epeirogenic sea-level. In the southern British Isles, the basin topography had a strong bearing on the location and style of sedimentation during this sea-level rise. As well as acting as local sediment sources, the highs also formed barriers between discrete, subsiding basins within which circulation was restricted. This allowed thick successions of organic-rich mudstones (Oxford Clay and Kimmeridge Clay) to accumulate in the basins under stagnant conditions. These became the main source rock for North Sea oil.

By the end of the Late Jurassic, most of the British Isles had become emergent once more (Plate 14), leaving a saline **lagoon** in the Wessex–Weald Basin, where marshes and soils periodically developed, intercalated with layers deposited during occasional marine incursions. The overall decrease in sea-level at this time was related to global climate change and to a new phase of uplift occurring to the west of the British Isles, associated with rifting in the Rockall Trough and the onset of active sea-floor spreading.

Throughout the whole of the Jurassic, Ireland was situated on its own separate emergent landmass that slowly encroached north-eastwards. As a result, the only Jurassic strata found in Ireland are limited to the eastern and north-eastern regions (Plates 12–14).

10.5.3 Cretaceous (51°–58° N, 142–65 Ma)

A very late Jurassic eustatic sea-level regression, along with a renewed period of tectonic activity, resulted in much of the British Isles rising above sea-level at the beginning of the Cretaceous. In southern England, the last Jurassic and the earliest Cretaceous strata were deposited under continental lagoonal, lake and fluvial conditions (e.g. the Portland Beds, Purbeck Beds, Hastings Beds and Weald Clay), while rifting to the west continued to cause uplift in south-west England (Plate 15). Further north and to the east of the Pennine landmass, the Lower Cretaceous is entirely marine in character, comprising the Hastings Beds to Upper Greensand and Gault (Plate 15).

Towards the end of the Early Cretaceous, the depositional environment began to change, as a marine transgression flooded across the whole of southern England, and eventually submerged the Market Weighton High (SE(44)8842). Sea-levels continued to rise throughout the Late Cretaceous, flooding much of the Early Cretaceous landmass (Plate 16), leaving only a few, isolated emergent areas. These areas were probably of a low relief, as there is very little evidence for siliciclastic sediment being shed from them at this time. The cause of this marine transgression can be linked to tectonic–eustatic processes, as the tectonic regime across the British Isles was changing to one of mild compression associated with the opening of the central Atlantic Ocean. This produced a series of gentle thrusts and folds, including those in the emergent Weald Basin (Figures 10.8 and 10.9).

Throughout the Late Cretaceous, the dominant rock type to be deposited was the Chalk. This is very fine-grained, rhythmically-bedded carbonate composed of minute skeletal plates of tiny marine algae (**coccoliths**) and **foraminifers**. Unlike some other chalks, the white chalk typically found across most of the British Isles (known as the Chalk with a capital 'C') did not form as a deep-water **ooze** or in very shallow waters, but has been shown to have formed at a depth of 100–600 m, under normal marine conditions. Within the Chalk are thin (mm–cm) bands of marl and dispersed nodules or bands of chert (commonly referred to as **flint**), which are oriented parallel to bedding. The flints are biogenic in origin, forming during diagenesis after the Chalk had been deposited. The silica required to form the flints is primarily derived from sponge **spicules** (spines), although a small amount may also come from siliceous **diatoms** and **radiolarians** (which are both **microfossils**). Initially, the siliceous sponge spicules accumulated randomly across the sea floor along with the Chalk, dissolving over time as a result of the alkaline environment within the calcium carbonate sediments. As the dissolved silica flowed through the Chalk, localized acidic conditions in discrete zones associated with decaying organic matter (e.g. in burrows or marine organisms) caused the silica to resolidify.

The end of the Cretaceous (which is also the end of the Mesozoic Era), was marked by a catastrophic event, with the mass extinction of many fossil groups including the dinosaurs and **ammonites**. The cause of this mass extinction is still hotly debated, although most geologists agree that the blame lies mainly with the climatic effects of a large meteorite impact made worse by the emission of massive volumes of sulfur dioxide and other gases from flood basalts at several sites across the globe.

10.5.4 Tertiary (40°–50° N, 65–2 Ma)

In the north-west of the British Isles, igneous activity was widespread during the Early Tertiary. It consisted of extensive flood basalts, dyke swarms and plutonic intrusions (Figure 10.3), associated with the initial opening of the Rockall Trough and the North Atlantic Ocean. This renewed period of tectonic activity resulted in all of the British Isles forming a large emergent landmass (Plate 17), with only the south-east corner of England periodically partially submerged. Here, the Tertiary sedimentary succession forms a relatively thin onshore continuation of thicker successions (≤3 km) found in the offshore basins (Plate 17). Meanwhile, in the far north-east of the British Isles, tectonic activity resulted in the Shetland Platform undergoing a significant amount of uplift (~200 m) and tilting. The rapid erosion of this new highland culminated in significant deltaic and **submarine fan** complexes building out to the east, filling the Viking and Central Grabens with thick sedimentary successions (Plate 17).

Within the North Sea, some Tertiary sedimentary deposits exhibit localized gentle folding and faulting. This contrasts with sedimentary rocks in the south-east of England, which are characterized by low-angle thrusting and gentle folding of Neogene age, forming a continuation of deformation that commenced during the Late Cretaceous. This folding influenced the location and type of sedimentation that occurred throughout the Tertiary, with the Hampshire and London Basins consisting of a mixture of shallow-water marine and non-marine sandstones and mudstones, with the basins separated by the now gently uplifted Weald, which was covered by marsh (Plate 17). Despite the general eustatic marine regression during the Late Tertiary (Figure 3.2), rapid subsidence of the North Sea Basin due to post-rifting thermal relaxation allowed a largely marine succession to accumulate.

Throughout the Tertiary, the fauna and flora found in basin sediments across the British Isles and mainland Europe are indicative of a subtropical climate, even though the region was at approximately the same temperate latitude as at present. This subtropical climate is primarily a reflection of higher global temperatures at that time.

10.6 SUMMARY

- The British Isles migrated northwards, initially as part of the supercontinent Pangea and then as part of Eurasia, starting at a low latitude in which an arid climate prevailed, to its present-day high latitude and temperate climate.
- The types of sedimentary successions that formed throughout the Younger Cover were significantly influenced by a combination of eustatic as well as epeirogenic sea-level changes. These sea-level changes were in turn affected by local and regional tectonic activity.
- The Permo-Carboniferous boundary is marked by a period of tectonic and igneous activity, which produced a series of half-graben extensional basins, along with the intrusion of several major shallow-level intrusive bodies. Several of these bodies (including the Whin Sill) were intruded into basins that were still subsiding as a result of thermal relaxation of the crust.
- The half-graben basins continued to develop throughout the Permian into the Early Cretaceous, with synchronous sedimentary deposition and extension allowing thick sedimentary deposits to form within these basins.
- Following on from the Permo-Carboniferous, the next major stage of igneous activity did not occur until the Mid-Jurassic, when rifting and extension resulted in the extrusion and intrusion of tholeiitic to alkali basaltic material in the northern North Sea. Pyroclastic eruptions of similar age also occurred to the south-west of the British Isles, again associated with extension and rifting.
- By the Late Cretaceous, extension and faulting stopped, allowing Upper Cretaceous sedimentary successions of uniform thickness to be deposited across the basins, which, associated with thermal subsidence, produced a series of sag-basin deposits.
- In the Tertiary, the onset of the Alpine Orogeny in southern Europe along with rifting of the North Atlantic resulted in reverse reactivation of Mesozoic extensional faults across the British Isles. The low levels of compression produced in the British Isles by this deformation event formed a series of low-angle thrusts, buckle and monoclinal folds.

Now try the following question to test your understanding of this Section and the Younger Cover.

Question 10.5 Figure 10.10 is a sketch map showing the timing of rifting and collisional events adjacent to the British Isles. What relationship if any, can you see between these events and the following episodes in the development of the Younger Cover:
(a) Mid-Jurassic volcanism and uplift in the North Sea;
(b) Mid-Cretaceous tectonic movements;
(c) change from extensional to compressional tectonics in southern England;
(d) Tertiary igneous activity;
(e) faulting and folding of the Younger Cover in southern England?

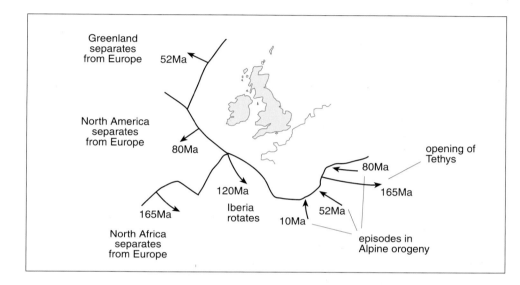

Figure 10.10 Sketch map showing the timing of continental separation and collision in regions adjacent to the British Isles. The arrows represent the direction of movement, the timing of which is indicated by the adjacent age. (From Dewey, J.F. (1982) *J. Geol. Soc.*, Vol. 139, Pt 4.)

11 THE QUATERNARY PERIOD

11.1 INTRODUCTION

Almost all the present-day landscape features of the British Isles were shaped during the Quaternary. When most people hear mention of the Quaternary, they think of it as an 'ice age'. However, study of land and marine sediments deposited during this time shows that the climate was in fact very varied. Sedimentary successions reveal a range from warm temperate to glacial conditions, with humid to semi-arid periods, involving a mixture of fluvial, aeolian, marine, **periglacial** (i.e. near-glacial, like the tundra) and true glacial deposits. In other words, the landscape attributed to this 'ice age' has actually been produced by a variety of processes during both glacial and **interglacial** periods.

In geological terms, the Quaternary is a very short time period and has only lasted for ~2 million years so far. This can be contrasted with the Tertiary and Carboniferous, which had durations of ~63 million years and ~64 million years respectively (Figure 2.1). Traditionally, the Quaternary is subdivided into two epochs:

- the Pleistocene (1.8 Ma–10 000 years ago); and
- the Holocene, representing the last 10 000 years of post-glacial time.

Although the currently accepted international base of the Quaternary (as defined by the International Union of the Geological Sciences, IUGS) is set at ~1.8 Ma, this may be amended in the near future to coincide with a major faunal change that has been recognized throughout north-west Europe at 2.5 Ma. This change correlates with a substantial ice rafting record in the North Atlantic Ocean, as well as the appearance of *Homo habilis* in Africa.

11.2 QUATERNARY DEPOSITS AND GEOLOGICAL MAPS

At present, the Earth is in the Holocene epoch, which many geologists regard as the latest interglacial episode. Deposits from this epoch and the whole of the Quaternary are referred to as **'drift'**, as they consist primarily of unconsolidated sedimentary deposits such as sands, gravels, clays and river alluvium, which have been left behind by receding ice-sheets. Careful examination of geological maps reveals that the only Quaternary strata shown on these maps belong to the Lower Pleistocene Crag deposits, a series of shelly sands and clays representing a transgression of the North Sea across East Anglia, prior to the main glaciations of the British Isles. Quaternary deposits are usually omitted from larger-scale geological maps of the British Isles, except for special 'Drift' editions.

11.3 HOW CAN THE QUATERNARY BE DEFINED?

Regardless of the short duration of the Quaternary, it stands out as a distinct time in the geological record, as it represents a period of considerable global climatic instability, which is quite different from the conditions during the Jurassic, Cretaceous or Tertiary. Evidence from fossil faunas and floras, and from estimates of past ocean-water temperatures, suggests that during the Mesozoic and Tertiary, the climate changed relatively slowly, with abrupt changes occurring rarely. In contrast, the Quaternary has been characterized by repeated cycles of rapid global climate change (Figures 3.2 and 11.1).

The Quaternary Period

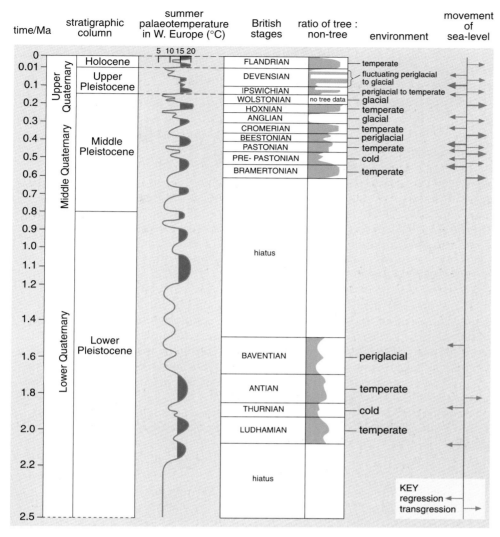

Figure 11.1 Changing palaeotemperature and floral diversity in the British Isles during the Pleistocene to Holocene. Palaeotemperatures are based on the change in oxygen-isotope signatures ($\delta^{18}O$). Warmer periods are characterized by a high abundance of trees, whereas cooler periods are dominated by grasses and shrubs (i.e. non-tree flora). (From *Geology of England and Wales, op. cit.*)

Although continuous successions of Tertiary and Quaternary sediments are found only on ocean floors and in a few very ancient lakes, an international research consortium has been correlating these with the fragmentary continental record, to obtain a more complete global picture of climate change during this time. A significant proportion of evidence for climate change has been obtained from the abundant assemblages of **benthonic** (bottom-dwelling) and **planktonic** (free-floating) microfossils found in cores of ocean sediments recovered during deep-sea drilling, many of which are particularly rich in the tiny calcareous skeletons of single-celled foraminifers (Figure 11.2). As many of these microfossil species are still living today, it is possible to estimate the water temperatures in which the organisms lived by using knowledge of their present-day ecology in association with the relative abundance of light and heavy oxygen isotopes in their skeletal calcium carbonate ($CaCO_3$). This can then be used to estimate how the global climate has changed. For example, modern day planktonic *Neogloboquadrina pachyderma* found at high latitudes in cold, icy seawater (e.g. around Greenland and Scandinavia) are coiled sinistrally (i.e. to the left), whereas samples of the same species living in lower latitude, warmer ice-free waters, coil dextrally (i.e. to the right, Figure 11.2).

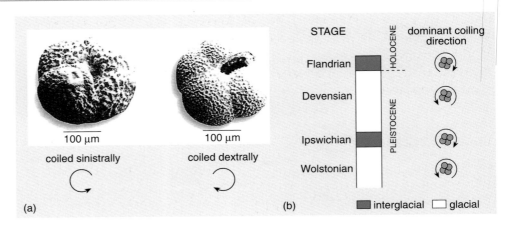

Figure 11.2 (a) Electron photomicrographs of two differently-coiled *Neogloboquadrina pachyderma* foraminifers coiling sinistrally and dextrally. (b) Relationship between interglacial and glacial stages and the dominant coiling direction of foraminifers found in a sediment core from the Norwegian Sea.

This and other similar lines of evidence (e.g. from pollen), have shown that towards the end of the Tertiary (~3 Ma ago), a series of alternating warmer and cooler climatic cycles occurred on a global scale, with a periodicity of ~100 000 years (Figures 3.2 and 11.1). These cycles are referred to as glacial–interglacial cycles and became much more pronounced 2.5 Ma ago, with the colder phases marked by expansions of glacier-ice on land and floating pack-ice in the oceans especially in the Northern Hemisphere. Over the last 2.5 million years, about 50 of these cycles have occurred, affecting conditions not only in the temperate latitudes, but also the tropical and subtropical regions (Figure 11.3).

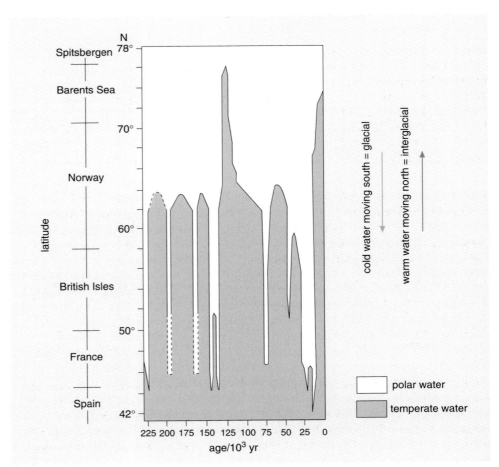

Figure 11.3 Changing surface seawater temperature in the North Atlantic during the past 225 000 years (nine glacial–interglacial cycles). Variations in seawater temperatures are based on the changing distribution patterns of planktonic foraminifers. During cold (glacial) periods, the low surface seawater temperature typical of high latitudes (e.g. Scandinavia, Barents Sea) extended down to the mid-latitudes (e.g. France, Spain).

The Quaternary ice age however, is not a unique event in geological history. Icehouse and greenhouse events have been recognized at several times in the more distant past.

Question 11.1 During which other geological periods have you encountered evidence of ice ages?

11.4 How the Quaternary began

Ice cover around the Earth's poles develops when the polar areas become partially isolated from the Earth's main atmospheric and oceanic circulation currents. For example, Antarctica has had a permanent ice cap since the Oligocene (38 Ma, an epoch in the Mid-Tertiary), after it split from the rest of Gondwana (Australasia, South America and India), and became separated from the main currents by the circum-Antarctic wind and ocean current systems. The kind of climatic pattern required for a *global* ice age however (which involves an unstable climate and extension of ice bodies into temperate latitudes), was not established across the Earth until the end of the Tertiary, when global tectonics also led to the partial isolation of the Arctic Ocean. Two important global tectonic readjustments that influenced this climate change were:

- the closure of the strait between North and South America, which resulted in the formation of a totally new ocean current system in the Atlantic, including the development of the Gulf Stream that keeps the climate in the British Isles so mild today;
- the shallowing of the Bering Strait between Alaska and Asia, which either limited, or at times of low sea-level, entirely cut off the circulation of water between the Arctic and Pacific Oceans. (During periods of extremely low sea-level, a land bridge emerged and allowed animals to cross between the two continents.)

In the Northern Hemisphere, the resulting deterioration of climate is first evident from the development of glacial deposits on Iceland at ~3.5 Ma, with material transported out from land by ice rafts and deposited offshore in North Atlantic sediments younger than 3 Ma. The first major continental glaciation of Europe dates from 2.4 Ma, when palaeobotanical evidence indicates that many temperate tree genera became locally extinct. It is uncertain when the highlands of the British Isles were first glaciated during the Quaternary, but this probably occurred over one million years before the first surviving glacial deposits (**tills**) were formed.

The oldest Quaternary deposits in northern Scotland are ice-deposited sediments, dated at ~0.8 Ma (Lower–Middle Pleistocene boundary). These are found on the continental shelf off the Aberdeenshire coast (i.e. to the east of the mainland). On land, the oldest deposits are found on the Shetlands and across the north-east of Scotland, dating from ~0.5 Ma (Middle Pleistocene).

Although glaciation processes most likely affected the north of the British Isles first, the oldest surviving deposits there are no older than the oldest equivalents found in the lowlands of England. The most probable explanation is that any older glacial deposits have been reworked and destroyed by successively younger events, with only the later deposits in each area standing any chance of preservation.

An indication of the severity of glaciation over a specific area can be obtained by looking at erosion-related features. For example, the size of **corries** (referred to as cirques in England and cwms in Wales) carved out by mountain glaciers, and the size of glacial ('U'-shaped) troughs cut by valley glaciers, can be used to estimate the amount of glacial and glaciofluvial erosion that has taken place

(Figure 11.4). Large corries and 'U'-shaped valleys indicate large amounts of erosion under glacial conditions, which may have been enhanced by mass movement of material under periglacial conditions. To produce large-scale features such as those found throughout Scotland, Ireland, northern England and Wales, implies long periods of ice cover.

(a)

(b)

Figure 11.4 (a) A typical well-developed corrie in Scotland.
(b) 'U'-shaped valley of glacial origin, Isle of Arran, Scotland. Note that the stream is a misfit river.

11.5 Glaciation of the British Isles

The base of the Quaternary in the British Isles has traditionally been placed at the base of the Red Crag deposits which are pre-Anglian in age (Figure 11.1), dating from the Upper Tertiary to Middle Pleistocene. These deposits are found across East Anglia, and typify cool temperate conditions and are known to be older than any of the glacial deposits in the British Isles. They are primarily marine in origin and contain **molluscs** and foraminifers, although some horizons also contain pollen assemblages and glacio-fluvial gravel. Studies of the molluscs and foraminifers suggest that the fauna are predominantly characteristic of Arctic waters, but were alternating with warmer-water fauna, indicative of a fluctuating climate with temperate and cold episodes (Figure 11.3).

Although the most spectacular evidence for glaciation in the British Isles comes from the mountainous regions, the sequence of repeated glacial and interglacial episodes can be worked out best in the lowlands, where large areas of the British Isles are still covered by glacial deposits (Figure 11.5). The distribution of these deposits gives a rough estimate of the maximum area of ice cover over the British Isles during the Quaternary. However, considerable local-scale erosion and removal of these deposits has occurred by subsequent glacial processes, and by much later human activities. In addition, some regions of the British Isles were not covered by glacial ice at the same time or during the same ice advance. This resulted in different regions experiencing different amounts of glacial erosion and/or deposition.

11.5.1 The three major ice advances

Evidence for only three major ice advances is known in the British Isles, the first of which began at about 0.45 Ma and is known as the Anglian ice sheet. There may have been earlier ice advances, associated with similar cold intervals during the earlier part of the Quaternary, such as those recognized in mainland Europe, but any evidence for these has been removed by the Anglian ice sheet. This sheet covered the whole of Scotland and Wales, with only the south of England and south-west of Ireland left ice-free. It also diverted the River Thames from a more northerly track across East Anglia to its present position. The Anglian advance was followed by a warm period and then another ice advance, the Wolstonian, possibly of similar extent at about 0.24 Ma. Glacial deposits left behind by these two advances are referred to as the Older Drift (Figure 11.5).

Although the eastern and southern margins of these two major ice sheets are well known, the western limits are less clear. The boundary to the west of Ireland can however be inferred from the location and nature of the continental shelf and the fact that at some point during the Quaternary, the whole of Ireland is known to have been glaciated at least once. Between the two Older Drift glacial events, it is probable that all rivers north of a line from the Thames to the Severn, at one time or another carried significant volumes of glacial melt-water, producing a series of large valleys, which are now occupied by much smaller **misfit rivers** (see Figure 11.4b), too small to have cut their valleys in their present state.

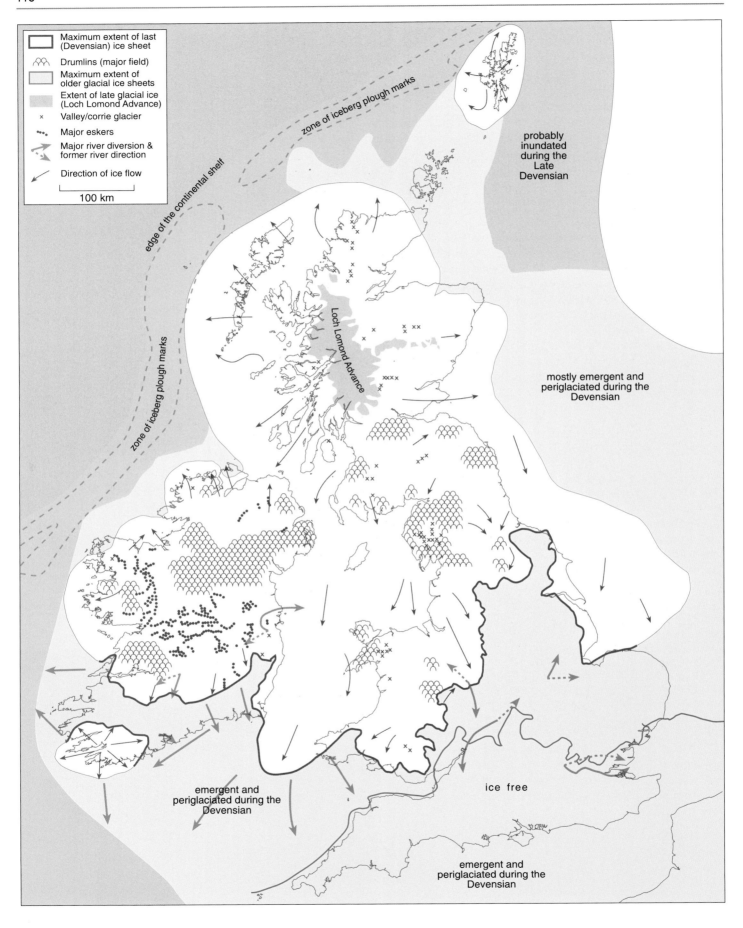

Figure 11.5   Palaeogeographic map revealing the distribution of glacial ice, gravels, drumlins and eskers (illustrated in Figure 11.6) in the British Isles, associated with the Devensian and Older Drift glacial ice sheets. The thick blue line represents the maximum extent of the last glaciation during the Devensian, delimiting the 'Newer Drift', while the thinner blue line represents the margin of older glaciations. The 'Older Drift' can only usually be recognized between the margins representing the maximum extents of the last and the older glaciations.

The third glacial advance, known as the Devensian, is the most recent major glacial stage to affect the British Isles. It lasted from 80 000 to 10 000 years ago, and reached its climax ~18 000 years ago (Figure 11.1). Glacial deposits from this period are referred to as Newer Drift, and consist partially of reworked Older Drift deposits. The maximum extent of the Devensian ice sheet can be seen on Figure 11.5, covering the majority of Scotland and Wales, leaving central and southern England, southern Wales and the south of Ireland (excluding the south-west corner, which had its own small ice sheet) ice-free but under periglacial conditions. The Devensian deposits are stratigraphically extremely complex due to the number of recessions and advances of widespread ice sheets and more localized ice lobes during this time. In the north-west of Scotland, Devensian ice flowed northwards over and beyond the Outer Hebrides, forming a series of ice-ploughed furrows still recognizable on the continental shelf today. To the north-east, the Orkneys remained unglaciated, whereas the Shetlands had their own small ice sheet. Recent work has shown that the British Isles and Scandinavian ice sheets were never in contact during the Devensian.

Most of the glacial landforms found throughout the British Isles such as the **esker** (Figure 11.6a) and **drumlin** fields (Figure 11.6b–c) across central England, Wales and much of Ireland, date from the Devensian. An esker is a long sinuous ridge of gravel, sand and silt, which is unrelated to the surrounding topography. As an ice sheet melts, it releases water and various glacial sediments into a series of discharge tunnels under the base of the ice. Sediment deposited as bed load is left behind as a ridge of sand and gravel after all the ice has melted. Drumlins form elongated egg-shaped mounds of unsorted boulders and glacial clay, extending up to 1–2 km in length and 6–60 m in height. No one really knows how they formed, but the most probable process is a rhythmic retreat of the ice sheet over newly glaciated land.

Another important impact of this event was the diversion of several major rivers such as the Derwent through East Yorkshire, the Severn between Wales and Bristol, the Liffey through Dublin and the Shannon north of Limerick (Figure 11.5).

By 13 000 years ago, the Devensian ice sheet had all but disappeared from the British Isles as global temperatures rose. The shrinking of the ice continued until 11 500 years ago, when a slight temperature decrease allowed it to build up once more, limited this time to the upland areas of Scotland, Wales and north-west England. This cold period lasted for ~1000 years and is referred to as the Loch Lomond Readvance (Figure 11.5).

(a)

(b)

(c)

Figure 11.6 (a) Series of eskers at Murrens on the border of Counties Meath and Westmeath, central Ireland.
(b) Drumlin field on Clare Island, Co. Mayo, Ireland.
(c) Satellite view of drumlin swarms in the valley regions of the Lake District and Pennines. The area shown is about 70 km across. The alignment of the long axes of the drumlins indicates that the direction of ice flow varied across the region. (British Geological Survey/NERC © 1992. All rights reserved.)

11.5.2 Glacial control on human distribution

Question 11.2 It can be seen from Figure 11.7 that the majority of palaeolithic (e.g. Old Stone Age) artefacts have been found south of the Devensian ice sheet, with these materials strongly concentrated in the south of England and northern France. What two opposing hypotheses can be suggested to explain this distribution?

A few artefacts have in fact been found within the southern margins of the Devensian ice sheet, so it would appear that the first hypothesis is the more likely, with Palaeolithic communities restricted to living in periglacial regions, where there were sufficient raw materials to sustain an existence. As the ice retreated, these communities could move northwards and develop new lands, and conversely, during periods of glacial advance, they would be restricted to more southerly regions.

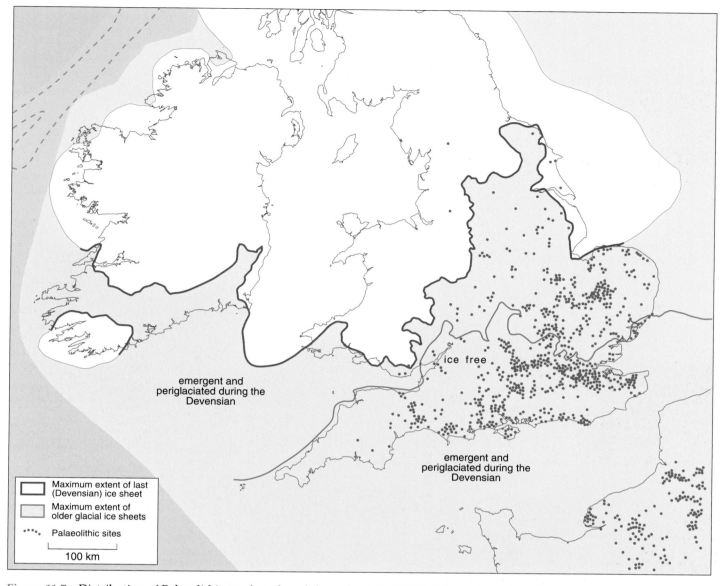

Figure 11.7   Distribution of Palaeolithic artefacts found throughout the British Isles and northern France.

Climate changes are not restricted to the pre-history of the British Isles. During the 14th–19th centuries, the local climate was cold enough to allow rivers as far south as the Thames to freeze over during the winter months. This period in history is referred to as the 'Little Ice Age', and demonstrates how the Quaternary climate is still fluctuating.

11.6 BETWEEN AND BEYOND THE GLACIATIONS

The Quaternary ice age and its glaciations had a much wider effect on the environment than just the land that was directly covered by ice. Both **permafrost** (the frozen layer of water below the surface) and a variety of periglacial processes influenced not just southern England and Ireland, but the whole of the British Isles immediately before and after each glacial event. For example, as the global climate grew colder, areas experienced periglacial conditions until the onset of glaciation. Then, as the temperature began to rise again, glaciation would give way to periglacial conditions before returning to a more temperate, interglacial state.

As well as affecting the environment, the glacial–interglacial periods also resulted in significant vertical crustal movements and fluctuations in sea-level. During glacial periods, landmasses covered by ice became isostatically depressed by the overlying mass of the ice. At the same time however, sea-levels fell even faster than the land, because of the volume of former seawater now locked up as ice. The net result was that the British Isles periodically formed a peninsula of mainland Europe during some of the more severe glacial periods. Ireland was still linked to Great Britain as recently as 14 000 years ago, and Great Britain did not become isolated from the mainland until less than 8000 years ago.

As the ice sheets retreated during interglacial periods, the release of this weight caused the British Isles to readjust isostatically, with maximum uplift occurring in the north and west where the thickest ice sheets had been. This in turn resulted in a scissors-like action, with the unglaciated south-eastern areas undergoing subsidence. At the same time as the land was rebounding, sea-levels began to rise as water from the melted ice sheet was released back into the seas. The combined effect of these vertical crustal movements and fluctuating sea-levels has left numerous examples of both raised beaches (Figure 11.8a) and submerged features (Figure 11.8b).

Figure 11.8 Evidence for sea-level change around the British Isles.
(a) A **raised beach** near Westward Ho! in Devon. Pebbles and boulders overlie a raised shoreline cut across inclined layers of slate forming the bedrock. The present-day shoreline is visible in the distance.
(b) Fossilized remains of a forest at Marros Sands, Pembrokeshire, south Wales. The trees, which grew during a period of lower sea-level, can now only be seen at low spring tides.

During interglacial periods, the vegetation was dominantly tundra-like, consisting of bog and open ground with discontinuous cover of Arctic herbs, grasses and dwarf shrubs. Woolly mammoth, reindeer, woolly rhino and musk ox roamed over this ground, and their fossil remains can be found in glacial deposits. Only in warmer periods could coniferous forests and then deciduous forests of oak, elm and ash colonize the British Isles, supporting badger, deer and fox, as well as more exotic species such as hippopotamus, lion and hyena that are now either extinct or confined to tropical areas.

Humans first appeared in the British Isles during the interglacial periods and eventually mastered the periglacial environment as hunters of mammoth, bison and reindeer. The oldest European **hominid** remains and authenticated stone tools date from ~0.5 Ma and include a human shin bone (tibia) and tooth excavated at Boxgrove, West Sussex in the mid-1990s. The most north-westerly Quaternary human remains in Europe date from an interglacial period, and were found in ~0.2 Ma sediments from Pont Newydd, north Wales. In addition, stone tools are frequently found in Quaternary lake sediments, river gravels, and occasionally as **erratics** in glacial deposits (Figure 11.9). Only in post-glacial times did humans develop the skills of crop and animal husbandry that ultimately permitted the development of urban civilizations. These skills were first developed in the Near East, but reached the British Isles with the first Neolithic (New Stone Age) people ~5500 years ago. With the development of modern civilizations, we have imposed our own ecosystems on the landscape, modifying many of the geological processes that sculpt them.

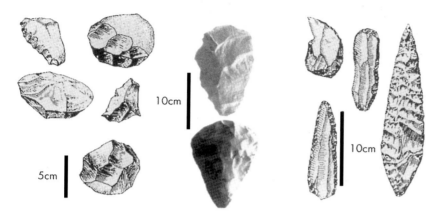

Figure 11.9   Palaeolithic stone axes and tools from 30 000 years ago.

11.7 Summary

- The Quaternary (~2 Ma to present), is characterized by a very varied climate, which has ranged from warm temperate to glacial conditions, along with humid to semi-arid periods.
- The oldest Quaternary deposits in the British Isles are ~0.8 Ma and consist of ice-deposited sediments on the continental shelf off Aberdeen. On land, the oldest sediments are 0.5 Ma, found in Scotland and the lowlands of England, representing the oldest surviving glacial deposits to be preserved in these areas.
- At least three major and several minor ice advances have affected the British Isles since the beginning of the Quaternary, with each glacial period lasting for 10–20 000 years, separated by even longer interglacial periods.
- Since the most recent retreat of the ice, both the land and sea-level have been rising as a result of isostatic readjustments.

12 Objectives for this book

We hope that your study of this book has given you a satisfactory taste of the really rather splendid, if intricate, geological history of the British Isles. We hope too that you have an appreciation of the complexities of some of the evidence upon which the geological history of the British Isles as presented in this book is based.

Now that you have completed this book, you should be able to:

1 Summarize and identify descriptions of the principal features of the main lithotectonic units of the British Isles, namely the *Precambrian Basement*, the *Caledonian Orogenic Belt*, the *Variscan Orogenic Belt*, the *Older Cover* and the *Younger Cover*.

2 Identify any of the main terranes making up the British Isles on the basis of a description of its age, main rock types, dominant structures, and plate tectonic setting.

3 Describe the differences between the Basement in the northern British Isles (north-west Scotland and northern Ireland) and the southern British Isles (England, Wales and southern Ireland).

4 Cite the evidence on which plate tectonic reconstruction of the evolution of the Caledonian Orogenic Belt is based.

5 Explain how the aftermath of the Caledonian Orogeny led to the formation of igneous plutons and the development of blocks and troughs.

6 Draw and recognize simplified palaeogeographic sketch maps of the British Isles representing the times when the Older Cover and the Younger Cover were being deposited.

7 Sketch or recognize graphic logs of the Carboniferous cycles of northern England and summarize the three modes of origin proposed for them.

8 Recognize and explain examples of different tectonic controls on sedimentation in the history of the British Isles, such as: accretionary prisms, block and basin sedimentation, basin and high sedimentation and inversion.

9 Summarize how plate tectonics, sea-level and climate change are interlinked, and recognize examples from this Book.

If you would like to know more, you could try an Open University course (see p. 4), or attend meetings or fieldtrips run by a local geological society. The Open University Geological Society, which welcomes non-students as well as students, has branches all over the British Isles; for information, see http://www.ougs.org/

Answers to Questions

Question 3.1

Over the last ~550 million years, the British Isles has formed a small portion of a series of different continental masses (being in fact part of two separate continents in the Cambrian), and has slowly drifted northwards to its present latitude.

Question 3.2

A crude correlation between these factors can be observed, with the sea-level low between 300–200 Ma corresponding to a time when all of the land was together in the supercontinent of Pangea (Figure 3.1e–f). The sea-level highs before and after this period correspond to times when the continents were drifting apart.

Question 3.3

There is a good correlation between the sea-level 'lows' and the Quaternary and Permo-Carboniferous ice ages, but global sea-levels were high during the Late Ordovician glaciation. This means that ice ages cannot be the only factor controlling sea-level changes.

Question 5.1

The Appendix shows that the onshore Younger Cover areas are much thinner (generally <1 km) than the offshore areas (which are up to 6 km thick). These offshore basins are of particular economic interest, as you will see when you examine the Younger Cover in more detail in Section 10.

Question 6.1

The oldest rocks in the British Isles are the Lewisian gneisses. They are located in the far north-west of Scotland, primarily to the west of the Moine Thrust Zone (MTZ) (Figure 5.2).

Question 6.2

The bulk of the Hebridean Terrane is made up from the undifferentiated Lewisian gneiss complex, cut through by a mafic–intermediate dyke swarm (Figure 6.6), and which is overlain by the unmetamorphosed Torridonian Sandstone.

Question 6.3

The most likely setting is at a destructive plate boundary, above a subduction zone (Section 4.2.2).

Question 7.1

The Moine Thrust Zone forms the north-western border of the Caledonian Orogenic Belt in Scotland.

Question 7.2

The sequence of basaltic pillow lavas followed by cherts and black shales is representative of a deep ocean basin, that is proximal to the spreading ridge and characterized by low rates of deposition. Moving up sequence, the occurrence of thick turbidites indicates that the environment of deposition has changed to a trench and trench slope region (see Section 4.2.2 *Sedimentary rocks* and *Structure*.)

Question 7.3

The oldest unit in the greywacke successions becomes progressively younger to the south-east. For example, moving south-eastwards between thrust slices ③ and ④, the oldest greywacke unit in thrust slice ③ is 18 million years older (six graptolite zones) than the oldest unit in thrust slice ④.

Question 7.4

Your completed section should resemble that in Figure A7.1.

Figure A7.1 Completed version of Figure 7.7.

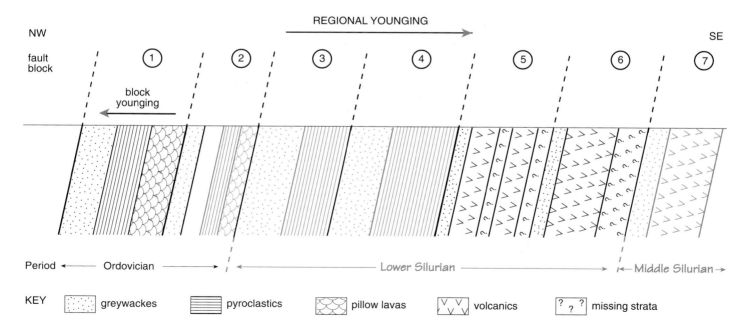

Question 7.5

All these features are characteristic of an accretionary prism that has developed above a subduction zone.

Question 7.6

This sequence represents oceanic trench sediments and the oceanic crust and upper mantle upon which they were deposited. As the oceanic crust and mantle has been thrust onto land rather than subducted, it is called an ophiolite (Section 4.2.4).

Question 7.7

As the granites crosscut the Caledonian structures, the granites must be younger.

Question 7.8

As the granites are not related to active subduction, they must have been generated as a result of continental collision, i.e. crustal thickening produced by collision resulted in the crust partially melting due to increasing temperatures.

Question 8.1

The Middle Old Red Sandstone/Devonian is found in two geographically distinct regions in the British Isles. The Middle ORS is limited to the north of Scotland (across the Moray and Pentland Firth areas (referred to as Caithness), the whole of the Orkneys and the southern Shetland Isles). The Middle Devonian is limited to north Devon in south-west England. It does not occur anywhere else in the British Isles.

Question 8.2

This absence indicates that either no sediments were deposited during this period, or that the sediments have subsequently been eroded away.

Question 8.3

Despite considerable local variation, repeating cycles of coarser sandstones and conglomerates (bed load sediments) followed by fine siltstones and mudstones (suspended sediments) can be recognized in all cases. Each cycle represents an overall fining-up sequence.

Question 8.4

The red coloration has been produced by the oxidation of iron from mafic minerals in the sediments and their source rocks. This is typical of sedimentary deposition in arid conditions.

Question 8.5

During the Early Devonian, extensive volcanic activity occurred across a number of regions in the Southern Uplands, the Midland Valley and the Central Highlands (e.g. Glen Coe in the Grampians). In the south-west of England, pillow lavas and pyroclastic deposits from the Early to Mid-Devonian are found among the sedimentary deposits. During the Late Devonian, volcanic activity was limited to the Orkneys and south Devon. All plutonic activity was limited to the Early Devonian, with major granites intruded in the Southern Uplands and Galway on one side, and northern England on the other side of the Iapetus Suture Zone.

Question 8.6

It implies that weathering and erosion must have happened quickly in order to include clasts of the eruptive units in sediments of similar age. This corroborates the fact that throughout the Devonian, the Caledonian mountains underwent rapid uplift and erosion.

Question 8.7

Overall, the Carboniferous strata are much more widespread across the British Isles than the Devonian strata.

Question 8.8

Sedimentation did not begin at the same time at all localities. Some areas have complete successions of the Lower Carboniferous, whereas in other areas the first Carboniferous sediments are later. For example, Carboniferous sedimentation begins over the Askrigg Block at 'time slice' 4 and over the Alston Block at 'time slice' 5, whereas the adjacent troughs contain successions 1–6.

Question 8.9

Granite plutons underlie many of the upland blocks including the south-east Southern Uplands, Alston and Askrigg Blocks.

Question 9.1

In the British Isles, deformation and metamorphism associated with the Variscan Orogeny was most strongly felt in south-west England (Devon and Cornwall), south Wales (Milford Haven to the southern edge of Glamorgan) and the south of Ireland (predominantly Counties Cork and Kerry).

Question 10.1

At the start of the Permian, the global sea-level was falling at an accelerating rate towards the Permian–Triassic boundary. An all-time sea-level low for the Phanerozoic occurred at about 245 Ma (Figure 3.2). Over the same time period, all of the continental landmasses had joined together to form one vast supercontinent, Pangea (Figure 3.1e–f).

After a slight fluctuation in the Triassic to Early Jurassic, from the Mid-Jurassic (~180 Ma) onwards, the global sea-level began to rise, reaching its peak level during the Late Cretaceous (~85 Ma, Figure 3.2). This rise in sea-level was accompanied by the break-up of Pangea into several smaller continents and the formation of a network of actively spreading oceans (Figure 3.1g–h).

Question 10.2

During the Early Carboniferous, a significant proportion of the British Isles was transgressed by a shallow, warm subtropical sea. The story is one of repeated prograding deltas. Initially, cycles of marine limestones, mudstones and sandstones were deposited, representing **prodeltas** and **delta fronts**. This was followed by cycles more characteristic of delta-top environments later in the Carboniferous.

Question 10.3

Based on the discussions in Section 3 and in this Section so far, you may have suggested the following mechanisms:

(a) *Glacially driven eustatic mechanism*: the formation and regression of major glacial systems will respectively lock up or release water into the ocean basins.

(b) *Tectonic–eustatic processes*: the formation of new spreading oceanic ridge systems uplifts the ocean basins, displacing water onto the land, producing a sea-level rise. The converse of this is the destruction of spreading ridge systems by continental collision, producing a sea-level fall.

(c) *Lithospheric extensional (basin formation) processes*: if subsidence is greater than thermal uplift during lithospheric extension, the subsiding basins can result in a localized or epeirogenic change in sea-level.

Question 10.4

On land, the Younger Cover is mostly limited to the south and east of England, with sedimentary strata generally ≤1 km thick, increasing to up to 3 km across the Dorset coast and Isle of Wight. In contrast, the strata are considerably thicker on the continental shelf areas (e.g. North Sea, Irish Sea, Western Approaches), varying between 1–5 km. Some mostly offshore areas of Younger Cover extend onto land in the north-east of Ireland, the Isle of Skye to the west of the Scottish mainland, and on the north-west coast of the Moray Firth, in the north-east of Scotland.

Question 10.5

The following events are related:

10 Ma	Africa collided with Europe (most recent episode in Alpine Orogeny)	Folding and faulting in southern England along reversely reactivated Mesozoic faults
52 Ma	Greenland and Europe separated	Tertiary igneous activity during rifting prior to ocean opening
80 Ma	Northern America and Europe separated	Change from extensional to compressional tectonics in southern England
120 Ma	Bay of Biscay opened (Iberia rotated)	Mid-Cretaceous tectonic movements
165 Ma	North Africa and northern America separated	Mid-Jurassic volcanism and uplift in the North Sea

Question 11.1

You already know about the tillites from the Precambrian Dalradian successions in Scotland (see Sections 3 and 6). In fact, at least four separate ice-age events are known to have occurred throughout the Precambrian on a global scale (Figure 3.2, Table 3.1). In addition, during the Phanerozoic, ice ages have also been recorded at various global locations in the Early Cambrian, the Ordovician–Silurian, the Devonian, and the Permo-Carboniferous (Table 3.1), with most of these lasting between 25–50 million years. It is interesting to note however, that there is no evidence for any icehouse events during the Mesozoic. At this time, the Earth was under constant greenhouse conditions.

Question 11.2

The distribution of palaeolithic artefacts may represent the original human distribution, or else that all palaeolithic sites north of the ice sheet's boundary may have been destroyed by subsequent glaciation processes.

Acknowledgements

Grateful acknowledgement is made to the following sources for permission to reproduce material in this book:

Figures 1.1, 7.5c John Watson/Open University; *Figure 3.1a–c* T.H. Torsvik *et al.* (1996) 'Continental break up and collision in the Neoproterozoic and Palaeozoic', *Earth Science Reviews*, **40**, Elsevier; *Figure 3.1d* T.C. Pharaoh (1999) 'Palaeozoic terranes and their lithospheric boundaries within the Trans-European Suture Zone (TESZ)', *Tectonophysics*, **314**, Elsevier; *Figure 3.1e* J. Golonka and D. Ford (2000) 'Pangean (Late Carboniferous–Middle Jurassic) paleoenvironment and lithofacies', *Palaeogeography, Palaeoclimatology, Palaeoecology*, **161**, Elsevier; *Figure 3.1f* P.J. Wylie (1970) *The Way the Earth Works*, Wiley, originally in R.S. Dietz and J.C. Holden (1970) *Jour. Geophys. Res.*, **75**; *Figure 3.1g, h* R.S. Dietz and J.C. Holden (1970) *Jour. Geophys. Res.*, **75** © 1970 American Geophysical Union; *Figure 3.2* L.A. Frankes, J.E. Francis and J.I. Syktus (1992) *Climate Modes of the Phanerozoic*, Cambridge University Press; *Figure 4.2 a(i)–c* from an illustration by Ian Worpole in P. Molnar (1986) 'The structure of mountain ranges', *Scientific American*, July; *Figure 6.1a–f* J.W. Schopf (1992) 'The oldest fossils and what they mean', in Schopf, J.W. (ed.) *Major Events in the History of Life*, Jones & Bartlett, courtesy J.W. Schopf; *Figure 6.2a* Donald R. Lowe; *Figure 6.2b* Andrew Knoll; *Figure 6.3* Chip Clark, National Museum of Natural History, Smithsonian Institution; *Figures 6.6a, 6.9d* Andy Tindle/Open University; *Figures 6.6b, 9.10b, 10.1a, b, 10.2b, 11.4a* Dave Rothery/Open University; *Figures 6.8, 8.3b, 8.4b, c, d, 8.11, 9.6d, 11.6b* Arlëne Hunter/Open University; *Figures 7.4, 7.8* W.S. McKerrow *et al.* (1977) *Nature*, **267**, pp. 237–239, copyright © 1977 Macmillan; *Figures 7.5b, 8.3a, 8.9a, 10.8* Ivan Finney; *Figures 7.9, 11.4b* Peter Sheldon/Open University; *Figures 7.10, 7.12* N. Woodcock and R. Strachan (2000) *Geological History of Britain and Ireland*, Blackwell; *Figures 8.2, 8.5* R. Anderton *et al.* (1979) *A Dynamic Stratigraphy of the British Isles*, Allen & Unwin; *Figure 8.8* M.R. Leeder and A.H. McMahon (1988) *Geology of England and Wales*, Geological Society; *Figure 8.10* G. Kelling and J.D. Collinson; *Figure 8.12* E.H. Francis (1979) 'British coalfields', *Science Progress*, **66**, Science Reviews; *Figure 9.3* P.A. Meere (1995) 'The structural evolution of the western Irish varicides', *Tectonophysics*, **246**, © 1995, Elsevier; *Figure 9.6a, b* Rodney Gayer; *Figure 9.6c* J.L. Roberts (1989) *The Macmillan Field Guide to Geological Structures*, Macmillan; *Figure 9.9b* Enterprise Oil; *Figures 10.2a, 10.9d* David Peacock; *Figure 10.2c* Angela Coe/Open University; *Figure 10.5* A.W. Woodland (1995) 'Petroleum and the Continental Shelf of NW Europe', Vol. 1, *Geology*, Institute of Petroleum, London; *Figure 11.2* Geoffrey Boulton; *Figure 11.3* B.G. Anderson and H.W. Borns (1994) *The Ice Age World*, Scandinavian University Press; *Figures 11.5, 11.7* J.C.W. Cope *et al.* (eds) (1992) *Atlas of Palaeogeography and Lithofacies*, Geological Society; *Figure 11.6a* Robbie Meeham; *Figure 11.8a* Derek Mottershead; *Figure 11.8b* Kevin Church/Open University; *Figure 11.9* S. Jones *et al.* (1992) *The Cambridge Encyclopedia of Human Evolution*, Cambridge University Press.

Every effort has been made to trace copyright owners, but if any have been inadvertently overlooked, the publishers will be pleased to make the necessary arrangements at the first opportunity.

Appendix

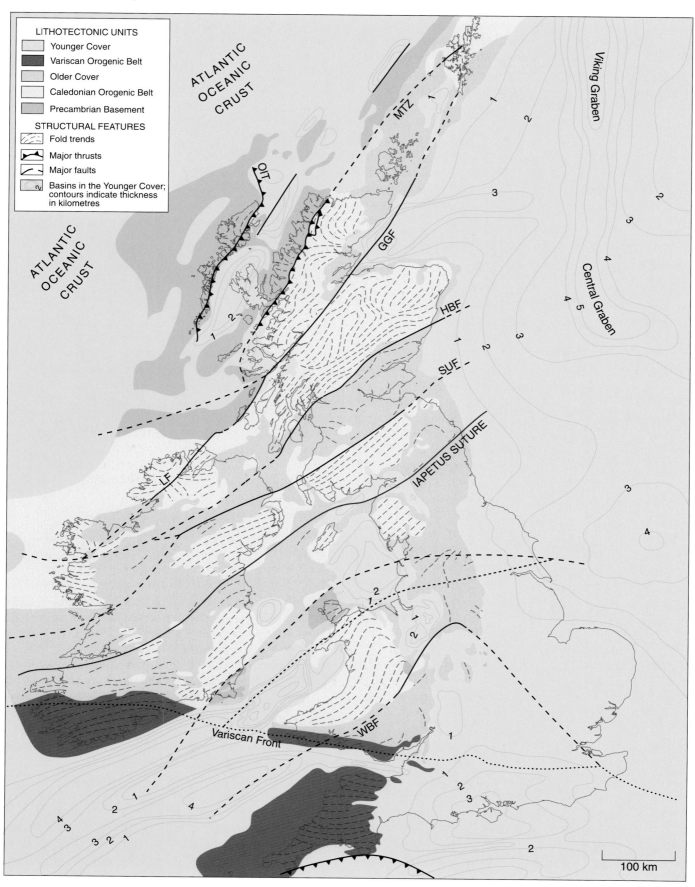

Glossary

accretionary prism A wedge of sediment that forms in an ocean trench directly above the shallowest part of a subduction zone.

active continental margin A continental margin that is also a plate boundary; typically a subduction zone.

active rifting A splitting apart of tectonic plates as a result of thinning; driven by mantle processes.

alkaline Containing a high concentration of alkalis (Na_2O, K_2O) relative to silica (SiO_2); in other words, silica-undersaturated.

alluvial fan Deposit in the form of a cone-shaped apron of sediment laid down where a river loaded with debris emerges from mountains onto a flat lowland plain.

ammonite A member of a subclass (Ammonoidea) of extinct cephalopods (a class of the Phylum Mollusca) with a spirally coiled external shell. The latter was divided internally into a series of chambers by transverse septa with elaborately folded margins, forming complex marks or sutures on the inner wall of the shell. The subclass ranged from the Devonian Period to the Cretaceous Period, though one order, the ammonites, was especially successful during the Jurassic and Cretaceous.

amphibolite A metamorphic schist or gneiss consisting predominantly of amphibole (pronounced 'am-fibb-o-lite').

amphibolite facies A medium-grade metamorphic facies in which hornblende and plagioclase are stable in rocks of mafic composition. Muscovite, kyanite or sillimanite may be stable in rocks of pelitic composition.

andesite A fine-grained, intermediate, volcanic rock with a chemical composition similar to that of microdiorite and diorite. Characteristic of ocean–continent destructive plate boundaries.

anhydrite An evaporite mineral, $CaSO_4$.

anthracite Coal with a high carbon and low volatile (gas) content.

anticline A fold that is convex upward (arch-like), or one that had such an attitude at some stage in its development.

arkose A sandstone type of sedimentary rock with a fragmental texture, mainly composed of quartz and feldspar, varying from sand- to gravel-sized grains; rocks generally accepted as arkose usually have a feldspar content greater than 25%.

asymmetrical fold A fold where one fold limb is steeper than the other and, as a consequence, the limbs are of unequal length.

back-arc-basin A miniature sea-floor spreading environment caused by extension behind a mature volcanic arc.

basalt A fine-grained, mafic, igneous rock with a chemical composition similar to that of dolerite and gabbro. Usually formed by the cooling of lava that has erupted at the Earth's surface, but it may also be found in minor intrusions such as sills and dykes, if cooled very quickly. On cooling, it may develop characteristic polygonal joint patterns.

basement Any rocks underlying a covering lithotectonic unit.

Basement A Precambrian and Lower Palaeozoic basement.

basin A syncline with inward-plunging ends.

batholith A set of related and mutually intrusive plutons.

benthonic Dwelling on the sea-bed.

berthierine A complex, sheet silicate mineral often found in ironstone deposits.

bifurcating Dividing into two.

biota Animals and plants.

bioturbation The churning of sediment by burrowing organisms, causing local disruption or destruction of original stratification. Intense bioturbation usually indicates well-oxygenated conditions and relatively slow rates of deposition.

bivalve A member of a class of molluscs (Bivalvia), with a two-valved shell; e.g. cockles. Most bivalve shells have bilateral symmetry, with each valve being a mirror image of the other (although *individual* valves lack symmetry). Some bivalve groups, such as oysters, have asymmetric shells.

blueschist facies A high-pressure, low-temperature metamorphic facies in which blue amphiboles (e.g. glaucophane) are stable in rocks of mafic composition.

brachiopod Member of a phylum (Brachiopoda) of sessile marine invertebrates having a two-valved shell. The shell is bilaterally symmetrical, but the valves are dissimilar. In articulate brachiopods, the two valves articulate posteriorly at a hinge. In inarticulate brachiopods, there is no hinge, the valves being held in position entirely by muscles. Brachiopods were more diverse in former times, especially in the Palaeozoic.

braided Referring to a river whose channel shows braiding; in other words, splits into a pattern of smaller channels, which rejoin and divide again.

breccia A coarse- or very coarse-grained fragmental rock, in which individual clasts are angular (pronounced 'bretch-ia').

buckle fold A fold formed by shortening along the length of the layers, usually not directly related to faulting.

build-up A mounded mass of limestone differing from surrounding deposits, that can be shown originally to have had positive relief on the sea-floor as it accumulated. Origins may vary, ranging from an *in situ* accumulation produced by sessile shelly benthos (as with modern tropical coral reefs) to piles of carbonate sediment emplaced by currents, often with assistance from sessile benthos trapping the sediment at the surface.

buried topography Ancient hills and valleys that have subsequently been buried by later sediments. The surface of the buried topography often forms an irregular unconformity.

calc-alkaline Rich in CaO and alkaline components (Na_2O, K_2O), with a low FeO/MgO ratio.

calcite A carbonate mineral with the chemical composition $CaCO_3$ (aragonite is the other calcium carbonate polymorph).

calcrete A fossil soil horizon formed in arid conditions, consisting of fine-grained, nodular to unbedded limestones that have precipitated around and between siliciclastic grains in the soil by evaporation processes.

caldera A larger (>1 km diameter) version of a volcanic crater, formed mainly by subsidence over the roof of a partly evacuated magma chamber.

caliche A fossil soil horizon formed in arid conditions, consisting of fine-grained, nodular to unbedded limestones that have precipitated around and between siliciclastic grains in the soil by evaporation processes.

carbonate platform A marine area of mostly shallow-water depth, where there is a thick accumulation of carbonate sediment. There are five different morphological types of carbonate platform: the carbonate ramp, rimmed shelf, epeiric platform, the isolated platform or bank, and the drowned platform.

chalk A very fine-grained limestone that is often quite soft and usually white or creamy in colour. Chalk is usually composed of minute skeletal plates (coccoliths) of tiny marine algae or forams (protozoa).

chert A fine-grained silica-rich rock formed by the biochemical precipitation of silica (SiO_2) from solution. May be found as irregular nodules (commonly referred to as flint) in chalk, or as a bedded deposit.

chevron fold A fold with straight limbs and a sharp hinge.

chronostratigraphic column A vertical succession of rock units based on their absolute ages.

clasts Mineral, rock or skeletal fragments found in most sedimentary rocks, and a few, notably pyroclastic, igneous rocks. They may range in size from small sand-sized grains to large boulders.

cleavage Planes of weakness in metamorphic rocks, notably slates, caused by the alignment of platy minerals.

coal A dense, black, carbon-rich deposit formed after the burial and compression of plant material accumulated as peat in stagnant swamps, marshes and bogs. It is usually so thoroughly altered that plant remains are only rarely preserved.

coccoliths Minute skeletal plates of tiny marine algae which, on death, have accumulated as a carbonate-rich mud on the sea floor to form, after compaction and burial, chalk.

conglomerate A fragmental sedimentary, coarse-grained rock, composed of fragments larger than 2 mm in diameter cemented in a finer-grained matrix; the consolidated equivalent of gravel.

constructive plate margin A plate margin where oceanic parts of adjacent plates diverge, and each is added to by sea-floor spreading. Because of the sense of relative motion between the adjacent plates, it is sometimes referred to as a divergent plate margin. Marked by a ridge on the ocean floor, and often referred to as a mid-ocean ridge.

continental shelf The edge of a mass of continental crust that has been thinned and stretched by rifting and is therefore sufficiently low-lying to be flooded by a shallow sea.

continental slope Relatively steep (between 1° and 15°, but typically about 4°) slope at the oceanward edge of a continental shelf, terminating at a depth of 1.5–3.5 km in a continental rise.

corrie A small, basin-shaped feature carved out of a mountain side by a glacier.

craton A rigid area of continental lithosphere that has been tectonically inactive or almost inactive for hundreds of millions of years.

crinoid A member of a class (Crinoidea) of echinoderms, with or without a stalk, having a cup-shaped body and moveable arms, with radial food grooves leading to the mouth; commonly known as sea-lilies. Crinoids were more diverse in former times, especially in the Palaeozoic.

cross-stratification Beds or laminae of siliciclastic or calcareous sediment that are inclined at an angle to the surface on which the sediment was deposited. Cross-stratification results from the preservation of the depositional slope of a migrating bedform.

crust The compositionally distinct layer overlying the mantle. In the terrestrial planets, the crust is richer in silica than the mantle. The Earth has two kinds of crust: oceanic crust and continental crust.

cyanobacteria A group of bacteria containing a blue, photosynthetic pigment, and formally regarded as (blue-green) algae.

delta The geomorphological feature that develops if a large volume of sediment builds out from the mouth of a river.

delta front The steeper seaward-dipping part of a delta, situated landward of the prodelta and seaward of the delta plain.

delta plain The top of a delta adjacent to the land, the delta plain is dominated by alluvial processes. The other two parts of a delta are the delta front and prodelta.

destructive plate margin A plate margin where the oceanic edge of a plate descends in a subduction zone below the oceanic or continental edge of an adjacent plate. Because of the sense of relative motion between the adjacent plates, this is sometimes called a convergent margin. Surface expressions include a trench on the ocean floor and volcanoes on the overriding plate, which form an island arc if this is oceanic.

diagenesis A term embracing all the changes that take place in sediments after deposition, and which influence the preservation of fossils, but which do not include changes that take place because burial is so deep that metamorphism occurs. Interactions between sediment and pore-waters are common during diagenesis, and often involve dissolution, precipitation and recrystallization. The growth of nodules within a sediment is a diagenetic phenomenon.

diatom A siliceous microfossil.

diorite A coarse-grained, intermediate, intrusive igneous rock with a chemical composition similar to that of andesite and microdiorite; formed as a result of slow cooling at depth below the Earth's surface.

dolomite Carbonate rock-forming mineral of composition $CaMg(CO_3)_2$. Occurs in limestone that has been exposed to magnesium-bearing solutions, notably in evaporitic settings. Often used to refer to carbonate rock composed predominantly of this mineral.

downthrow The relative downward displacement on one side of a fault.

drift Unconsolidated sedimentary deposits, such as sand, gravel, clay and river alluvium that have been left behind by receding ice-sheets after a period of glaciation.

drumlin An elongated, egg-shaped mound of unsorted boulders and glacial clay, moulded by the rhythmic retreat of an ice-sheet over newly glaciated land.

ductile The opposite of brittle. Behaviour of solids that respond to pressure by flowing or folding rather than by fracturing.

duplex A series of thrusts stacked one on top of the other.

dyke A sheet-like igneous body that is intruded nearly vertically into existing rocks and that cuts across the boundaries between those rocks.

dyke swarm A number of associated dykes, often paralleling one another. A group of dykes radiating from a common centre (radial dykes) is described as a radial dyke swarm.

echinoid Member of a class (Echinoidea) of vagrant echinoderms, lacking arms or a stalk, but possessing moveable spines; commonly known as sea-urchins.

effusive (eruption) Volcanic eruption in which material is erupted mainly in the form of lava flows.

Eon The largest unit of geological time. Earth history is divided into two Eons: Cryptozoic ('hidden life') and Phanerozoic ('visible life').

epeirogenic A change in sea-level due to local events such as tectonic movements.

ephemeral Short-lived.

Era A large unit of geological time, based on fossil characteristics in sedimentary rocks. The Cryptozoic Eon is divided into three Eras: Hadean, Archean and Proterozoic. The Phanerozoic Eon is divided into three Eras: Palaeozoic, Mesozoic and Cenozoic.

erosion The process of wearing away the surface of the Earth's crust, usually by the mechanical action of water or ice, or by particles transported by wind, water or ice.

erratic An 'exotic' boulder deposited by an ice-sheet, far away from the place where it originated.

esker A long, sinuous ridge of gravel, sand and silt deposited at the base of a melting ice sheet.

eustatic A change in sea-level affecting the whole globe.

evaporite A sedimentary deposit formed by chemical precipitation in hot, dry conditions as evaporation occurs in areas of restricted water supply. Any remaining water becomes saturated with ions such as Na^+, K^+, Cl^- and SO_4^{2-}, and minerals such as halite (rock salt, NaCl) and gypsum ($CaSO_4.2H_2O$) may accumulate.

exhumation The ascent of a solid rock towards the surface, usually by uplift accompanied by erosion. Note that exhumation will only be the same as uplift if erosion keeps pace with uplift.

exotic terrane A piece of crust or terrane that has originated at a distance from the rocks that now lie adjacent to it.

explosive (eruption) Volcanic eruption in which material is mainly flung out explosively by the force of escaping volatiles.

extrusive Term describing igneous rocks that are the result of volcanic eruption at the Earth's surface. This includes both lavas and pyroclastic rocks.

facies (metamorphic) A term that characterizes a metamorphic rock by its temperature and pressure of formation.

fault Fracture in a rock along which there has been an observable amount of displacement. Faults can range from a displacement of a few centimetres to fractures of continental proportions. Faults are rarely single surfaces; normally, they occur as parallel or sub-parallel sets of surfaces along which movement has taken place.

faunal assemblage A particular collection of animal species.

feldspar A framework silicate mineral containing sodium, potassium and/or calcium, that is found in many igneous rocks in the Earth's crust.

felsic rock Igneous rock, e.g. granite, with an SiO_2 content of 66 to 75% (by weight), usually more than 10% quartz, and less than 20% mafic minerals and alkali feldspars.

fining up A grading upwards from coarse-grained at the base to fine-grained at the top.

flash flood A sudden rush of water down a river or a watercourse.

flint See **chert**.

flood basalt A series of large-volume basaltic flows (typically of the order of 10^6 km^3 in total volume), best known in continents but also found on oceanic crust. Flood basalt provinces occur sporadically, and although related to mantle plumes, their cause is not well understood.

footwall The body of rock underneath a dipping fault plane.

foraminifer A member of a major group of marine, single-celled organisms that secrete a (usually) tiny, multi-chambered calcareous shell. Both benthic and planktonic forms are common. Often abbreviated to 'forams', they are abundant as microfossils. Planktonic forms, in particular, are widely used for stratigraphical correlation.

fossil Remains or impression of an organism preserved in the geological record. The two most widely recognized kinds of fossil are body fossils and trace fossils.

fractional crystallization The physical separation of crystals from the magma in which they occur (i.e. from a partial melt). Because the first crystals to form are less rich in silica than the remaining magma, fractional crystallization leads magma to become enriched in silica.

gabbro A coarse-grained, mafic, intrusive igneous rock with a chemical composition similar to that of basalt and dolerite; formed as the result of slow cooling at depth beneath the Earth's surface.

glauconite A bright green sheet silicate mineral of the mica group, formed in marine sediments during, or soon after, their deposition (in other words, diagenetically).

gneiss A medium- or coarse-grained metamorphic rock, composed of alternating bands of felsic and mafic minerals, formed at very high pressures and temperatures during regional metamorphism.

goethite An iron hydroxide mineral, FeO.OH.

graben A structure formed when a block is downthrown between two parallel (or near-parallel) normal faults.

graded bedding Sedimentary beds in which there is a vertical gradation in grain size; in most instances, this is from coarser-grained at the base of the bed to finer-grained at the top.

granite A coarse-grained, felsic, intrusive igneous rock with a chemical composition similar to that of rhyolite.

granodiorite A coarse-grained felsic, igneous rock in which plagioclase forms over 67% of the total feldspar content.

granulite facies The highest grade metamorphic facies, in which orthopyroxene is stable in rocks of mafic composition. Orthopyroxene and kyanite or sillimanite may be stable in rocks of pelitic composition.

graphic log Largely pictorial diagram of a sedimentary succession based on observations of bed thickness, lithology including texture (especially grain size), sedimentary structures, fossils and palaeocurrents.

graptolite A member of an extinct (Palaeozoic) group of colonial animals, distantly related to chordates. Most were pelagic.

greenhouse A period in the Earth's geological history when the mean global temperature was higher than today's mean temperature.

greensand A sandstone containing a high abundance of the bright green mineral glauconite.

greenschist facies A low-grade metamorphic facies in which actinolite (a green amphibole), epidote (a green chain silicate), and chlorite (a green sheet silicate) are all stable in rocks of mafic composition. Muscovite is stable in rocks of pelitic composition.

greywacke A texturally immature sedimentary rock containing larger grains in a fine-grained matrix of clay- and silt-sized particles. The larger grains may range from sand- to gravel-sized particles and are composed of quartz, rock fragments and feldspar. In a greywacke, the matrix materials should constitute more than 15% by volume.

Glossary

grid reference A pair of rectangular co-ordinates used for point location based on the National Grid.

gypsum An evaporite mineral, $CaSO_4.2H_2O$.

halite A mineral with the formula NaCl and purely ionic bonding. Also known as rock salt.

hangingwall The body of rock above a dipping fault plane.

hominid Relating to humans.

horst The structure formed when a block of country is upthrown relative to the surrounding country between two parallel (or near-parallel) normal faults.

icehouse A period in the Earth's geological history when the mean global temperature was lower than today's mean temperature.

igneous rocks Rocks formed by the cooling and crystallization of magma (*ignis* is Latin for 'fire'). Usually characterized by an interlocking crystalline texture, with crystal size dependent mainly on the rate of cooling. Other textures are shown by very rapidly chilled igneous rocks with a glassy texture, and pyroclastic rocks with a fragmental texture.

inlier Outcrop of older rocks completely surrounded by younger strata.

interglacial A time period between two glaciations.

intermediate rock An igneous rock, such as diorite, which has an SiO_2 content of 52 to 66% (by weight). Usually, rocks of this type have 0 to 20% quartz, significant proportions of mafic minerals, and plagioclase feldspar that is about equally rich in Na and Ca.

intrusive A term describing igneous rocks formed by the cooling and crystallization of magma beneath the Earth's surface. Coarse-grained intrusive rocks, such as granite and gabbro, must have cooled slowly; they are also known as plutonic rocks.

inversion tectonics A reversal of a tectonic regime due to crustal deformation and shortening.

ironstone A clay-rich, sedimentary rock with a high percentage of iron; generally nodular in form, but sometimes oolitic.

island arc An arcuate belt of volcanoes, typically forming volcanic islands near the leading edge of the overriding plate at a destructive plate boundary.

isoclinal folds A fold with an interlimb angle of 0°; rather like a hairpin.

isostasy The tendency for regions of crust to sit at their neutrally buoyant level.

isostatic compensation The situation when isostasy has enabled regions of crust to reach their neutrally buoyant level. This is common on the Earth because the weakness of the asthenosphere allows it to flow to compensate for sagging or upwarping of the mantle part of the lithosphere.

isotope One of two or more atoms with the same number of protons but a different number of neutrons in the nucleus.

lagoon A shallow-water protected area between the continent and a barrier (e.g. a barrier island or carbonate build-up). A lagoon is protected from major coastal currents and high-amplitude waves, and is therefore a generally low-energy environment, except during storms, when the barrier may be breached. Lagoons occur in both carbonate platforms and siliciclastic environments.

lava The term used to describe magma when it is flowing at the surface, or the rock formed when this solidifies.

lava flow Lava emplaced in a single effusive episode.

limestone A sedimentary rock usually made of the mineral calcite ($CaCO_3$) but occasionally including an Mg-rich variety known as dolomite ($(Ca,Mg)(CO_3)_2$). Limestones may be biological in origin or may result from direct chemical precipitation of calcite from freshwater or, more often, from seawater.

lithosphere Outer, rigid shell of a planetary body, which in the Earth's case is broken into a number of tectonic plates. Consists of the crust and upper mantle.

lithostratigraphic column A vertical succession of rock units.

lithotectonic unit A unit of rock with a distinct geological history based on lithology and tectonic structures, but not correlated with any particular geological period.

load cast A sedimentary structure formed when a coarse-grained sediment, such a sand or silt, is deposited on top of soft, incoherent mud. The coarser sediment sinks unevenly into the underlying mud, expelling any pore water, thus allowing compaction to take place; can vary in shape from a slight bulge to a deeply rounded mass.

mafic rock Igneous rock, e.g. gabbro, with an SiO_2 content of 45 to 52% (by weight), containing no quartz, but rich in mafic minerals and Ca-rich plagioclase.

magma Molten rock, generally containing suspended crystals and dissolved gases. When extruded onto the surface, it is usually known as lava.

magma chamber A volume within the crust that is occupied by magma. If magma solidifies in a magma chamber, it forms a pluton.

mantle The silicate part of the Earth (or other terrestrial planet) that surrounds the core and is ultramafic in composition. Although almost entirely solid, that part of the mantle below the lithosphere is capable of convection.

mantle plume A cylindrical upwelling from deep within the mantle, which is the cause of a hot spot.

mantle wedge The mantle (wedge-shaped in cross-section) belonging to the overriding plate above the subducting slab at a destructive plate boundary.

marble A metamorphic rock formed from pure (monomineralic) limestone.

marl A carbonate-rich mudstone.

mélange A body of rock composed of disrupted fragments (a metre to a kilometre in scale) of pre-existing rock bodies.

metamorphic rock Rock whose texture and/or mineralogy has been changed by the action of heat and/or pressure (usually, both are involved). Metamorphic rocks can be derived from sedimentary rocks, igneous rocks or pre-existing metamorphic rocks.

metamorphism Changes brought about in a rock as a result of recrystallization in the solid state (i.e. without melting) owing to an increase in pressure or temperature, or both.

metasediment A sedimentary rock that has undergone metamorphism.

metazoa Multicellular organisms made up of different organs and tissue types.

micrite Microcrystalline calcite produced either biologically or by precipitation from seawater; also referred to as lime mud.

microfossil A body fossil of minute size, best studied under a microscope, although microfossils are also often visible under a hand lens.

mis-fit river A river that is too small to have cut the valley through which it currently flows.

molasse A poorly sorted mixture of sands and conglomerates produced by the rapid erosion of newly uplifted mountains.

molluscs (formally Mollusca) Phylum containing three large classes, the bivalves (Bivalvia), cephalopods (Cephalopoda) and gastropods (Gastropoda), as well as some minor groups such as chitons. Molluscs generally have a calcareous dorsal shell that grows incrementally, and is secreted by mantle tissue. They have a mantle cavity containing gills, and a body with a dorsal visceral mass, beneath which is a muscular foot, much modified in cephalopods.

monocline A one-limbed fold.

MORB Shorthand for mid-ocean ridge basalt.

mudstone A very fine-grained sedimentary rock containing grains less than 62.5 µm in size; called shale if it is fissile (in other words, splits easily into thin layers). The term mudstone encompasses siltstones (grains 4 to 62.5 µm in size), claystones (grains less than 4 µm in size) and those rocks containing a mixture of both clay- and silt-sized particles.

nappe A body of rock that has undergone considerable horizontal transport in an orogenic belt; maybe an asymmetrical fold with a sub-horizontal axial surface.

normal fault A steeply inclined fault in which the movement is predominantly dip-slip, and in which the fault plane dips towards the downthrown block.

obduction The process by which slices of oceanic lithosphere are forced upwards onto adjacent continental crust in a collision zone.

oceanic crust The crust making up the Earth's ocean floors. It is of mafic chemical composition, and is thus poorer in silica than the continental crust.

ooids Spherical grains found in limestones. They resemble fish roe. Today, they are mostly formed by the chemical precipitation of aragonite (a metastable polymorph of calcium carbonate) in agitated water in shallow marine areas. Sediments containing ooids are said to be oolitic; on compaction and burial, they may form an oolitic limestone. In ancient ooids, the aragonite has been replaced by more-stable calcite.

oolitic limestone A limestone consisting of small rounded grains produced by chemical precipitation in agitated water in a shallow, marine environment.

ooze A fine-grained, marine mud consisting of the skeletons of tiny planktonic organisms.

ophiolite A slice of oceanic crust and upper mantle that has escaped subduction and become emplaced onto the continental crust in a collision zone.

orogenic belt A structural complex that forms a mountain chain or a former mountain chain. Typically, continental lithosphere has been shortened horizontally and thickened vertically by major thrust faults and large-scale folds with cleavages.

outcrop The area across which a particular rock unit would otherwise be visible if all the superficial deposits, soil, vegetation and buildings were removed.

outlier An outcrop of younger rocks separated from the main outcrop, and surrounded entirely by older strata.

paired metamorphic belt The association of two metamorphic belts, one characterized by high pressure and low temperature, and the other by relatively low pressure and high temperature.

palaeocurrent The current direction that prevailed at the time when a succession of sediments was deposited. Palaeocurrent directions can be determined from features such as cross-stratification and flute marks.

palaeomagnetic Relating to ancient geomagnetic fields.

partial melting The phenomenon in which (at any particular pressure) successive parts of a rock melt over a range of temperatures, because different minerals have different melting points. The more silica-rich minerals begin to melt at lower temperatures, so the first melt to form is richer in silica than the average composition of the starting material.

passive continental margin A continental margin, originally formed by continental splitting, that was left behind as sea-floor spreading carried it, and the other parts of the oceanic crust bordering it, away from the active spreading ridge.

passive rifting A splitting apart of tectonic plates as a result of thinning due to lateral extensional processes.

peridotite A dense, coarse-grained, ultramafic rock with a crystalline texture, characteristic of the lowermost oceanic crust and the principal rock type forming the upper mantle. Peridotite is composed largely of olivine and pyroxene.

periglacial Near-glacial, tundra-like.

Period A unit of geological time based largely on the fossil characteristics in its sedimentary rocks (a subdivision of Era). Most geological Periods span several tens of millions of years.

permafrost A frozen layer of water below the ground surface; found in periglacial conditions.

permeability A measure of the flow rate of a fluid through a porous rock or sediment.

pillow lava The distinctive morphology of basaltic lava erupted under water at low effusion rates.

planar stratification Layers of sediment deposited as a flat carpet over a surface.

planktonic The descriptive term for water-borne organisms that float or swim weakly, generally drifting with currents. The term includes phytoplankton (largely algae) and zooplankton (animals and other single-celled organisms). Zooplankton can be microscopic, (e.g. larvae of benthic animals) or macroscopic (e.g. jellyfish).

plate A part of the Earth's lithosphere that is surrounded by plate boundaries, and which behaves (for most purposes) as a fragment of rigid shell.

plate margin A boundary between adjacent tectonic plates. Plate margins may be conservative, constructive, or destructive.

pluton A large body of igneous rock (up to about 30 km across) intruded into the crust at sufficient depth for the groundmass to be coarse-grained.

pop-up basin A basin formed by strong rotation and uplift of a previously subsiding rifted trough and sag-basin as a result of crustal shortening.

porosity A measure of the percentage of the bulk volume of a rock or sediment occupied by pore space.

Principle of Faunal Succession The principle, first recognized by William Smith, that there is a recognizable succession of fossils for each period of geological history, and that this sequence is the same wherever the rocks of that period occur.

Principle of Superposition The principle of stratigraphy enunciated by William Smith, stating that if one series of rocks lies above another then the upper series was formed after the lower series, unless it can be shown that the beds have been inverted by tectonic action.

prodelta The furthest offshore portion of a delta, which is situated at the distal edge of the delta front and marked by slow, fine-grained deposition.

progradation The building out of sediment in a distal direction; e.g. the seaward migration of a delta, and the building out in a distal direction of an alluvial fan or point bar.

pull-apart basin An extensional basin, usually of relatively small size, formed in a strike-slip zone.

pyroclastic rock A fragmental volcanic rock formed by explosive eruption.

pyroxenite A very dark, coarse-grained ultramafic rock consisting entirely of the mineral pyroxene.

quartzite A metamorphic rock formed from pure (monomineralic) quartz sandstone.

radial dyke One of a series of dykes arranged in a radial pattern around a central pluton.

radial dykes (radial dyke swarm) Dykes radiating away from a volcanic conduit or magma chamber.

radiolarian A siliceous microfossil.

radiometric dating The dating of rock by the energy released during radioactive decay. Uranium, thorium and potassium have the only isotopes that produce a significant amount of radiogenic heat in the Earth today.

raised beach Beach deposits stranded at altitude by a fall in relative sea-level.

recumbent folds A fold with an almost horizontal (dip < 10°) axial plane.

red beds Sedimentary rocks that are reddish-brown due to coatings of iron (Fe^{3+}) oxide, and which may form post-depositionally. Red beds are generally taken as indicative of a highly oxidizing environment of deposition.

relief (topographical) Variations in the elevation of the land surface; e.g. hills and valleys.

reverse fault A steeply inclined fault in which the movement is predominantly dip-slip, and in which the fault plane dips towards the upthrown block. Reverse-type faults that display a low inclination from the horizontal are called thrusts.

rhyolite A fine-grained, felsic, igneous rock with a composition similar to microgranite and granite.

ridge (ocean) A submarine mountain range found in the middle of the ocean, where two tectonic plates are moving apart at a constructive plate margin.

ring dyke A thick dyke with a roughly circular outcrop pattern above a magma chamber that dips steeply outwards. Subsidence along the fault represented by a ring dyke is the main way in which a volcano caldera develops.

rock units Divisions of a lithostratigraphic column; groups, formations, members and beds.

sag-basin A basin-like structure that results from subsidence due to the weight of sediment along with thermal relaxation of the crust.

salt dome A folded structure, formed by the upward movement, under the influence of gravity, of a highly ductile salt deposit.

sandstone A rock composed of sand-sized grains (62.5 µm to 2 mm), where the grains are often mainly quartz but may also include significant amounts of rock fragments and feldspar. Sandstones composed of quartz and feldspar are termed arkoses. Smaller amounts of mineral oxides, bioclasts and clay minerals may also be present. Sandstones may have a quartz or calcite cement or a clay matrix.

schist A medium- or coarse-grained metamorphic rock with a large proportion of platy minerals, such as mica, which are aligned in one direction, defining an often undulating foliation. Fairly high temperatures (>400 °C) and pressures are required for its formation.

sea-floor spreading The process whereby the oceanic lithosphere is added to at a constructive plate margin.

sedimentary rocks Rocks formed from sediments that have undergone changes such as compaction and cementation to turn them into rock.

shale A mudstone that is fissile; in other words, it easily splits into thin layers.

sheeted dykes The layer in the oceanic crust above the gabbro but below the lavas, consisting of nothing but dolerite dykes.

siderite An iron carbonate mineral, $FeCO_3$.

siliciclastic The term used to describe a sediment or sedimentary rock composed predominantly of the silicate mineral residues of weathering, such as rock fragments, quartz, other unweathered silicate minerals, and clay minerals.

sill A generally horizontal sheet-like intrusion that is mostly concordant with the bedding of the strata that it intrudes.

siltstone A very fine-grained sedimentary rock containing grains 4 to 62.5 µm in size.

sinistral fault A strike-slip fault where the sense of movement of the block viewed across the fault plane is to the left.

slab The term used to describe the inclined, downgoing part of an oceanic plate at a destructive plate boundary.

slate A fine-grained, low-grade metamorphic rock formed by the recrystallization of a mudstone during regional metamorphism. It is characterized by very closely spaced, flat foliation planes caused by the alignment of platy minerals, and along which it may be split very easily; this is often referred to as slaty cleavage.

species The lowest unit in the taxonomic hierarchy: a population(s), of genetically related individuals sharing a common evolutionary history, and, in sexually reproducing organisms, capable of interbreeding to produce viable and fertile offspring.

spicules The silica-rich spines of a marine sponge.

stratigraphic column The array of geological time units (Eons, Eras and Periods) that results from stacking them vertically, with the oldest at the base overlain by successively younger units.

strike-slip fault A fault in which a sideways (in other words, horizontal) slip is the dominant sense of movement, either to the right (dextral) or left (sinistral). You may see strike-slip faults referred to elsewhere as transcurrent, wrench or tear faults.

stromatolite A finely layered, often crinkled or lumpy looking accumulation of carbonate muds or sands, trapped by surface-dwelling mats of filamentous photosynthetic bacteria or algae.

subduction The process whereby the oceanic part of a plate descends at an angle below another plate at a destructive plate margin.

submarine fan A thick, cone-shaped mass of sediment deposited at the base of the continental slope by the action of turbidity currents, or the mass movement of sediment down the slope due to instability caused by the pressure of accumulated sediments of local earthquakes.

superterrane A series of small terranes faulted into one main region.

suture zone The trace of a defunct destructive plate margin where continent–continent collision has occurred. Marked by a belt of mountains and usually including a few slivers of oceanic crust (ophiolites) that escaped subduction with the rest of the vanished ocean floor.

syncline A fold that is convex downward (in other words, U-shaped), or one that had such an attitude at some stage in its development.

syn-rift sedimentation Sedimentation occurring at the same time as extensional faulting.

tectonic Relating to structures produced by deformation.

terrane A piece of crust, defined by clear boundaries, that differs significantly in its tectonic evolution from neighbouring regions.

thermal relaxation The sinking of crustal material as it cools and becomes less buoyant after a period of igneous activity.

tholeiite A silica-oversaturated basalt.

thrust A special type of reverse fault where the fault plane has a dip not more than 45° from the horizontal (typically less than 20°).

till (tillite) A clay-rich glacial deposit left behind by receding ice sheets.

tip-related fold A fold that develops as a result of deformation at the tip of a thrust fault.

trace fossil A fossil recording the activities of an organism, such as burrows, tracks, borings, droppings or bite marks.

transform fault A fault on the ocean floor linking two offset stretches of a constructive plate margin, allowing the adjacent plates to slide sideways past each other. It is therefore a type of conservative plate margin.

transgression An advance of the sea over the land.

trench A deep depression (typically about 8 km deep) on the ocean floor, marking the site where an oceanic plate descends below an adjacent plate at a destructive plate margin.

trilobite A member of an extinct class of arthropods (Trilobita), having a body divided longitudinally into three lobes. The body is also divided into a head, central trunk and tail. They were confined to the Palaeozoic Era.

turbidite A succession of beds deposited by a turbidity current.

turbidity current A density current in which a mixture of dense turbid sediment and water flows downslope beneath surrounding, less dense water.

type area A locality that exhibits the best example of a particular geological feature or features.

unconformity A geological boundary representing a break in the deposition of sediments, or erosion between formation of rocks below and deposition of rocks above the unconformity surface.

vent A point from which a volcanic eruption occurs.

volcaniclastic Fragmental texture in a rock of volcanic origin.

volcanic rocks The products of volcanic eruption.

wadi A steep-sided, narrow valley scoured in desert mountainous areas by the water and debris resulting from torrential intermittent rains.

zone fossil A fossil limited to a specific, short period of geological time, and therefore useful for the relative dating of rocks.

INDEX

accretionary prism 19–21, 49–50, 56–59, 126
active continental margin 18, 40
active rifting 14, 16–17
alkaline (rocks) 18, 52–53, 100, 103, 108, 110
alluvial fans 63–64, 106
Alpine Orogeny 103–104, 110–111, 127
Alps 9
Alston Block 69–72, 94, 126
amagmatic 14
Amitsôq Gneiss 30
amphibolite 19, 34–36
Anglian ice sheet 117
Anglo-Brabant landmass 70–72, 75–77
anhydrite 106–107
Antarctica 11, 115
anthracite 76
Appin Group 42, 44–45
Arctic Ocean 115
Argyll Group 42, 44–45
Armorica 34, 80–81
Armorican Orogeny 61
Askrigg Block 69–72, 126
Atlantic Ocean 9, 11, 13, 17, 80, 95, 99–100, 102–104, 109–110, 112, 114–115
Avalonia 10, 34, 39, 50–51, 53–58, 60, 66, 78, 81
Avalon–Midland Platform Terrane 24–25, 39, 52–53, 58

back-arc basins 52–53, 56, 60, 72, 96
Badcallian Event 35–36
Ballantrae ophiolite 22, 50, 56, 59
Baltica 80–81
Barren Red Series 76
Basement 23, 26–27, 29–30, 32–35, 37, 39–40, 42–43, 51
basement 23, 26, 30, 42
basin and high 102
Ben Nevis 67
Ben Vuirich Granite 46
benthonic 113
Bering Strait 115
berthierine 107
bivalves 66
blocks 67, 70–71, 74, 78–79, 90, 126
blueschist facies 19–20
Bodmin Moor 82
Borrowdale Volcanic Group 52, 54, 56
Bowland Trough 69–71
Boxgrove 122
brachiopods 66, 72
braided rivers 107
Bristol Channel basin 103
British Caledonides 41–42, 51
British Isles (definition) 5
buckle folds 104, 110
build-ups 53, 66

calc-alkaline (rocks) 18–19, 24, 35–36, 52–54, 56, 60
calcite 106
calcrete 64–65
caldera subsidence 67
Caledonian Orogenic Belt 9, 14, 23, 26–27, 29–31, 34, 39, 41–42, 61, 63, 66, 124–127
Caledonian Orogeny 10, 23, 27, 33–34, 39–41, 58, 61, 67, 71–72, 78, 81, 90, 124
Caledonides 29, 42, 51, 57
caliche 65
Cambrian 27, 33, 37–38, 41, 44, 51–53, 56, 80, 125, 127
carbonate build-ups 53, 66
carbonate platform 66, 76, 81
Carboniferous 7–8, 11, 27–29, 58, 61–62, 68–72, 74–75, 78, 80–82, 84, 88–93, 95–96, 101–102, 124, 127
Carboniferous Limestone 62, 69, 129
Celtic Sea Basin 100, 103, 108
Central Graben 100, 103, 110
Central Highlands Terrane 24–25, 34, 42, 44, 46, 49, 59, 126
Chalk 98, 105, 109
chamosite 107
Charnia Supergroup 32
chert 20, 24, 47, 49, 59, 81–82, 109, 125
chevron folds 86
chronostratigraphic 8
cirque 115
Cleveland Dyke 99–100
Climacograptus 48
Clyde Plateau Lavas 68–69
coal 28, 73–74, 76–78
Coal Measures 62, 74–76, 79
coccoliths 109
conservative plate margin 14,
constructive plate margin 14, 55, 99
continental rifting 15
corals 72
corries 115–116, 118
Crag 112
Craven Fault 71
Cretaceous 11, 13, 95, 98–99, 102–104, 108–110, 126
crinoids 53, 72
Cryptozoic 8, 30–31, 34
Culm Basin 70
cwm 115
cyanobacteria 32

Dalradian Supergroups 34, 42, 44–46, 51, 59
Dartmoor 27, 82
delta 66, 73–76, 97, 108, 110, 126
destructive plate margin 14, 18, 21, 125
Devensian 113, 118–121

Devonian 10, 27–29, 34, 41, 50, 57–58, 60–61, 63–68, 78, 80, 83, 88, 126–127
diatoms 109
Didymograptus 48
Dinantian 69–70, 74
diorite 18
Diplograptus 48
dolomite 69, 95, 106
drift 28, 112
drumlin fields 118–120
duplex systems 85, 87
Durness Limestone 39
dyke swarm 24, 35, 109, 125

East African Rift 15
East Fife volcanics 69
echinoids 66
Ediacara fauna 32–33
epeirogenic sea-level changes 9, 72, 74, 78, 98, 107–108, 110
ephemeral lakes 65, 96
ephemeral vents 69
esker 118–120
Eurasia 11, 110
eustatic sea-level changes 9, 12–13, 71–74, 78, 81, 98, 108, 110, 126
evaporites 51, 95–96, 99, 106
evaporite successions 106–107
exotic terrane 21, 24, 26, 29, 33

faunal assemblages 20, 55
flash floods 64–65
flint 109
flood basalts 17, 99, 109
footwall ramp 84
foraminifers 109, 113–114, 117
fossil forest 122

gabbro 17
Galway–Mayo–Donegal High 67
gas reservoirs 106
gasfields 76–77
Gault Clay 109
geological time–scale 8
glacial–interglacial cycles 114
glacial-ice 98, 114
glauconite 98
Glen Coe 67
gneiss 24, 34–35, 38, 43, 83, 125
goethite 107
Gondwana 9–11, 20, 34, 51, 71–72, 80–82, 94–95, 98, 115
graben 15, 72, 91, 105
Grampian Group 42, 44–45
Grampian Orogeny 43–44, 46, 49, 56, 59
granite 18, 21, 27, 29, 41–43, 46–47, 58, 65–67, 71–72, 81–82, 87, 102, 126
granodiorite 34
granulite facies 19–20, 36

graptolites 47–49, 125
Great Glen Fault 25–26, 42, 44, 46, 57, 67, 78
Great Oolite 108
greenhouse 51, 115
greenschist facies 19
greensand 98, 105
greywacke 47, 49, 58–59, 66, 83, 125
Gulf Stream 95, 115
gypsum 106–107

half-graben 72, 101, 103–104, 110
halite 106–107
Hampshire Basin 110
Hastings Beds 109
Hebridean Terrane 24–25, 27, 34, 37, 39, 44, 46, 125
Hercynian Orogeny 61
Highland Boundary Fault 25–26, 29, 42, 44, 46, 57, 65, 67, 78, 126
Himalayas 9, 66
Hogs Back 104
Holocene 112
hominid 122
Homo habilis 112
horst 15

Iapetus Ocean 9–10, 39, 41, 44–46, 49, 51, 53–57, 59, 67, 78, 81
Iapetus Suture 10, 34, 57–59, 67, 129
Iberia 80–81, 127
ice rafting 115
icehouse 51, 115
Iceland 115
Inferior Oolite 108
interglacial 112, 117
International Union of the Geological Sciences (IUGS) 112
inversion tectonics 90–91, 94, 103, 105
Irish Sea Platform 53–54, 56–57
ironstones 107
island arcs 18
Isle of Man 53, 104
isostatic compensation 67, 71–72, 84, 103, 108, 122–123

Jurassic 10, 77, 95, 97–99, 101, 107–108, 110, 126

Kimmeridge Clay 97, 102, 105, 108
Knoydart Orogeny 43–44, 46

Laurasia 11
Laurentia 10–11, 34, 39, 46, 50–51, 56–60, 66, 78, 80–82
Laxfordian Event 35, 40
Leinster Granite 63
Leinster–Lakes Terrane 24–25, 41, 52–54, 58, 60

Lewisian 24, 27, 31, 34–35, 38, 42–44
lithospheric extension 14, 72, 76, 78, 80, 93, 98, 100, 126
lithostratigraphic 8, 42
lithotectonic units 23, 30–31
'Little Ice Age' 121
Lizard ophiolite 22, 27, 66, 80–81, 83–85, 87, 125, 130
Lizard Terrane 24–25
load casts 48
Loch Lomond Readvance 118–119
London Basin 110
Lothian Volcanics 69
Lower Carboniferous 69–70
Lulworth Crumple 105

Magnesian Limestone 106
Market Weighton High 102, 109
marl 106–107, 109
mélange 24
Mendips 102, 107
metamorphic Caledonides 26–27, 29, 31
metasediment 30
metazoans 32
micritic limestone 76
microfossils 109, 113
Mid North Sea High 99, 108
Midland Platform 25, 52, 54, 56–57
Midland Valley Sill 91–92
Midland Valley Terrane 24–25, 29, 46
Millstone Grit 62, 74–76
misfit rivers 116–117
Moine Supergroup 27, 34, 39–40, 42–44, 59
Moine Thrust 24–27, 29, 34, 39–40, 42–44, 56–57, 125
molasse 21
Monian Terrane 24–26, 39, 41, 52–53
monocline 104–105, 110
Monograptus 48
Monograptus turriculatus 48
Moray Firth Graben 100, 103
MORB 50, 59, 66

Namurian 62, 69–70, 74, 76, 88–89
nappe 21
Nemagraptus 48
Neogene 110
Neogloboquadrina pachyderma 113–114
Neolithic 123
New Red Sandstone 96
New Stone Age 123
Newer Drift 119
non-metamorphic Caledonides 26, 29, 31
North Sea 58, 93–94, 97–100, 102–103, 106–110, 112, 127
North Sea oil 108
Northern Highlands Terrane 24–25, 27, 34, 39, 42–44, 46, 49, 59
Northumberland Trough 69–72, 91

obduction 22
Ochil Hills 67–68
Oilfields 77, 106–107
Old Red Sandstone (ORS) 61–66, 78, 81, 94–96, 126
Older Cover 23, 27–29, 31, 61, 80–86, 90, 94, 103, 106
Older Drift 117, 119
Oligocene 101, 115
ooids 107
oolitic limestone 97, 105, 107–108
ophiolites 18, 22, 24, 59, 80, 87, 126
Orcadian Basin 61, 65
Ordovician 10, 37–38, 43, 47–56, 59–60, 80, 101, 127
Oxford Clay 97, 102, 105, 108

Pacific Ocean 115
paired metamorphic belt 19–20
Palaeolithic artefacts 120–121, 127
palaeomagnetic data 14, 20
Pangea 9, 11, 13, 81, 95–96, 98–99, 107, 110, 125–126
passive continental margin 16–18, 44–45, 55
passive rifting 14, 16–17
patch reefs 53
Pennines 5, 31, 58, 62, 74, 96, 120
peridotite 17, 50, 130
periglacial 112, 116
permafrost 121
Permian 7, 28, 82, 88, 91, 93, 95–96, 98, 101–104, 106, 110, 126, 131
pillow lavas 17, 24, 47, 50, 66, 81–82, 125–126
planktonic 113
plate tectonics 14
Pleistocene 12, 95, 112–113
pop-up basins 90, 94
Portland Beds 102, 105, 109
post-orogenic granites 58–60, 67, 71, 82, 90
post-rift deposition 105
Precambrian and Lower Palaeozoic Basement 23, 26, 29–31, 34, 37, 42, 44, 55
Principle of Faunal Succession 8
Principle of Superposition 8
pull-apart basins 67, 88–89
Purbeck Beds 102, 105, 109
Purbeck–Isle of Wight Monocline 104
pyroxenite 34

Quaternary 28, 95, 112–123, 125

radial dykes 100
radiolarians 109
raised beach 122
red beds 24
Red Crag 117

red marl 106
Red Sea rift 15
reverse reactivation 90, 94
Rheic Ocean 9–11, 51, 66, 71–72, 78, 80–82, 90, 94, 98
rhyolite 18, 53, 60, 65
rift–shoulder unconformity 42
ring dykes 67, 100
Rockall Trough 108–109
Rotliegende 106

sag-basins 72, 79, 88, 90, 103, 110
Salisbury Crags 17, 68–69
salt domes 100
schist 24, 44, 83, 125
Scourian Event 35–36
Scourie dykes 35–36
serpentinite 85
Shap Granite 58–59, 66
sheeted dykes 17
Shetland Platform 110
siderite 107
Sidlaw Hills 67
Silesian 69
Silurian 39, 41, 46–50, 52–54, 58–59, 80–81
Snowdonia Volcanic Group 54, 56
Southern Graben 103
Southern Highlands Group 42, 44
Southern Uplands Block 70, 75
Southern Uplands Terrane 24–25, 46–47, 58–59
sponge spicules 109
'spring back' 98, 104
St. George's Land 70
Stainmore Trough 69–72
Staithes 97, 102, 107

Start Point 27, 83
Stephanian 62
stratigraphic column 8
stromatolites 32–33, 44–45
Stublick Fault 71
submarine fans 110
superterrane 24, 39, 46
suture zone 15, 21–22, 58, 94
syn-rift sedimentation 101

Tayvallich Volcanic Series 44–46, 59
terrane 23–24, 33–34, 39, 41, 58, 80–81
Tertiary 11, 95, 99–100, 103–104, 109–110, 115, 127
Tethys ocean 11, 98, 106–107
texturally immature (rock) 37
thermal relaxation 53, 72, 78–79, 84, 90, 103, 110
tholeiitic (rocks) 18, 24, 44, 52–53, 56, 60, 100, 103, 110
thrust tip-related folds 84
tillite 45, 51, 127
tills 115
tip-related folds 84
Torridonian 24, 27, 35–38, 125
Tournaisian 62
transform faults 14
Triassic 11, 28, 98, 101, 105, 126
trilobites 58, 66
troughs 67, 70–72, 74, 78–79, 90, 126
turbidites 20, 24, 47, 49, 53–54, 81–83
turbidity current 20, 48, 74
type area 35

Upper Greensand 109
U-shaped valley 115, 116

Vale of Eden 6, 95, 101
Vale of Moreton 102
Variscan Foreland 90
Variscan Front 25, 28, 83–85, 88–89, 94
Variscan Orogenic Belt 9, 14, 23, 27–29, 31, 61, 72, 80, 82
Variscan Orogeny 17, 23, 27, 41, 61, 71, 76, 80–82, 86–87, 89–91, 94–95, 98, 102, 126
Viking Graben 100, 103, 110
Viséan 62
volcanic front 54–55

wadi deposits 96
Weald Basin 109
Weald Clay 109
Weald Dome 103
Welsh Basin 53–54, 56–58, 60, 70
Wenlock and Ludlow Series 53
Wessex–Weald Basin 108
Western Approaches Basin 100, 103, 127
Westphalian 62, 69–70, 74, 77, 80, 82, 88–89
Whin Sill 91–92, 110
Wolstonian ice advance 117

Yoredale cycles 74
Younger Cover 23, 28–29, 61–62, 76, 80, 95, 101, 103–104, 106, 110–111, 125, 127

Zechstein Sea 106
zone fossils 48